T0291043

Advances in Deep Learning for Medical Image Analysis

Advances in Deep Learning for Medical Image Analysis

Edited by
Archana Mire, Vinayak Elangovan, and Shailaja Patil

CRC Press
Taylor & Francis Group
Boca Raton London New York

CRC Press is an imprint of the
Taylor & Francis Group, an **informa** business

First edition published 2022
by CRC Press
6000 Broken Sound Parkway NW, Suite 300, Boca Raton, FL 33487-2742

and by CRC Press
4 Park Square, Milton Park, Abingdon, Oxon, OX14 4RN

CRC Press is an imprint of Taylor & Francis Group, LLC

© 2022 Taylor & Francis Group, LLC

© 2022 selection and editorial matter, Archana Mire, Vinayak Elangovan, and Shailaja Patil; individual chapters, the contributors

ISBN: 9781032137162 (hbk)
ISBN: 9781032137193 (pbk)
ISBN: 9781003230540 (ebk)

DOI: 10.1201/9781003230540

Typeset in Times
by Newgen Publishing UK

To my husband Mr. Pradeep Thakare for his continuous encouragement and support.

Dr. Archana Mire

Contents

Preface

Deep learning (DL) is a powerful set of techniques for machine learning using artificial neural networks. Over recent years, we have witnessed DL revolutionizing all kinds of medical imaging, including X-ray, ultrasound, computed tomography, magnetic resonance imaging, and positron emission tomography. Numerous artificial intelligence-based tools have been developed to automate medical image analysis tackling problems such as recognition, segmentation, and analysis of various regions of interest, e.g., lesions, tumors, clots, blockages.

DL models have the ability to learn patterns and relationships in medical images, utilizing huge volumes of medical images. This book aims to promote the latest cutting-edge DL-driven research in medical image analysis. It demonstrates the validation of DL techniques on a range of digital medical images, including identification and classification of various diseases, such as heart disease, cancer detection, white blood cell classification, diabetes detection, etc.

With the present notions, we hope that the book will inspire all those who are interested in DL courses and want to apply it to medical image analysis.

Editor Biographies

Dr. Archana Mire

Dr. Archana Mire is currently working as Head of Department, Computer Engineering, Terna Engineering College, Navi Mumbai, India. She completed her Ph.D. from Sardar Vallabhbhai National Institute of Technology (SVNIT), Surat on Passive Image Forensic Techniques. She has more than 15 years of teaching and reseach experience. Her research interests include image processing and machine learning. She has published many articles in prestigious journals and conferences. She has published one Indian patent. She has also served as a review board member, editorial board member, and a program committee member for a number of conferences and journals. She is currently developing software 'ferroalloys calculation' for the steel industry.

Dr. Vinayak Elangovan

Dr. Vinayak Elangovan is currently the Program Chair of Computer Science at the Penn State University at Abington. He earned his Ph.D. in Computer Information Systems Engineering at Tennessee State University in 2014. He has 15 years of research and teaching experience. His research interests include computer vision, machine vision, robotics, multi-sensor data fusion and activity sequence analysis, and he has a keen interest in the development of software applications. He has published more than 20 peer-reviewed scientific articles in the field of artificial intelligence (AI). He has worked on a number of funded projects related to Department of Defense and Department of Homeland Security applications. He is active in the scientific community as a peer reviewer for highly acclaimed journals. He has also served as a review board member, editorial board member, and program committee member for a number of journals and conferences in the field of AI.

Dr. Shailaja Patil

Dr. Shailaja Patil is Dean (R&D) and Professor in the Department of Electronics and Telecommunications at Rajarshi Shahu College of Engineering, Pune, India. She has 30 years of teaching experience. She completed her Ph.D. from SVNIT with research focusing on "Statistical Techniques in Localization and Tracking Using Wireless Sensor Network." Her Ph.D. candidates are pursuing research in the internet of things (IoT), 5G networks, and image processing. She has more than 80 publications in peer-reviewed journals/conferences. She has delivered expert talks on wireless sensor network (WSN), IoT, software-defined networking (SDN), research methodology, and intellectual property rights (IPR) at various workshops, and presented a tutorial on SDN at the Institute of Electrical and Electronics Engineers (IEEE) international conference in Lisbon, Portugal. She has authored various book chapters and books. She has received research funding from Savitribai Phule Pune University (SPPU), All India Council for Technical Education (AICTE), and Department for Science and Technology (DST) for Research in WSN and IoT. She is a Fellow of the Institution of Engineers and a senior member of the IEEE. She is currently Chair of IEEE Women in Engineering (WiE), Pune section. She also holds a Postgraduate Diploma in IPR from Nalsar Law University, where she has trained in dealing with different aspects of IPR. She is a registered Indian patent agent. She guides students and faculty members in filing and prosecution of IPR. She has two granted patents to her credit, eight filed patents, and two copyrights. Recently she has been honored by AICTE with "Visvesaraya Best Teacher Award 2020" and by the Pune section of IEEE with "Best Researcher/Innovator of the Year Award 2020."

Contributors

Javeria Amin
Department of Computer Science,
  University of Wah
Wah Cantt, Pakistan

Sridhar P. Arjunan
SRM Institute of Science and
  Technology
Tamil Nadu, India

Alka Barhatte
MIT World Peace University
Pune, India

Mainak Chakraborty
Defence Institute of Advanced
  Technology (DIAT)
Girinagar. Pune, India

Manisha Dale
MES's College of Engineering
Pune, India

Abhishek Das
ITER, Siksha 'O' Anusandhan
Bhubaneswar, Odisha, India

Anusmita Das
Gauhati University
Guwahati, Assam, India

Kaushik Dehingia
Gauhati University
Guwahati, Assam, India

Sunita Vikrant Dhavale
Defence Institute of Advanced
  Technology (DIAT)
Girinagar, Pune, INDIA

Rajesh Ghongade
Bharati Vidyapeeth
College of Engineering
Pune, India

Jitendra Ingole
Smt. Kashibai Navale Medical College
  and General Hospital
Narhe, Pune, INDIA

Mdi B. Jeelani
Imam Muhammad ibn Saud Islamic
  University
Riyadh, Saudi Arabia

Snehal K. Joshi
Dolat-Usha Institute of Applied
  Sciences
Veer Narmad South Gujarat University
Gujarat, India

Vishnu Kumar Kaliappan
KPR Institute of Engineering and
  Technology
Coimbatore, Tamil Nadu, India

B. Lalitha
KPR Institute of Engineering and
  Technology
Coimbatore, Tamil Nadu, India

M. Menaka
Health, Safety and Environment
Indira Gandhi Centre for Atomic
  Research
Kalpakkam, India

Mihir Narayan Mohanty
ITER, Siksha 'O' Anusandhan
Bhubaneswar, Odisha, India

C. Muthamizhchelvan
SRM Institute of Science and
  Technology
Tamil Nadu, India

P. Pandiyan
KPR Institute of Engineering and
  Technology
Coimbatore, Tamil Nadu, India

S. Srinivasulu Raju
VR Siddhartha Engineering College
Vijayawada, Andhra Pradesh, India

Dania Sadaf
Department of Computer Science,
  COMSATS University Islamabad
Wah Campus, Wah Cantt, Pakistan

M. Saranya
Department of Electronics and
  Instrumentation Engineering
SRM Institute of Science and
  Technology
Tamil Nadu, India

Muhammad Sharif
Department of Computer Science,
  COMSATS University Islamabad
Wah Campus, Wah Cantt, Pakistan

K. A. Sunitha
SRM Institute of Science and
  Technology
Tamil Nadu, India

Rajasekaran Thangaraj
KPR Institute of Engineering and
  Technology
Coimbatore, Tamil Nadu, India

B. Venkatraman
Resource Management and Public
  Awareness Group
Indira Gandhi Centre for Atomic
  Research
Kalpakkam, India

Mussarat Yasmin
Department of Computer Science,
  COMSATS University Islamabad
Wah Campus, Wah Cantt, Pakistan

1 ANFIS-Based Cardiac Arrhythmia Classification

Alka Barhatte
MIT World Peace University, Pune, India

Manisha Dale
MES's College of Engineering, Pune, India

Rajesh Ghongade
Bharati Vidyapeeth College of Engineering, Pune, India

1.1 INTRODUCTION

The electrocardiogram (ECG) is a cardiac signal representing the recording of the electrical activity of the heart. Information such as heart rate, rhythm, and morphology in the form of conduction disturbances can be extracted from the ECG signal. The significance of the ECG is notable in that coronary heart diseases are major causes of mortality worldwide. The ECG varies between different individuals, due to the anatomy of the heart, and differences in size, position, age, etc. Thus, the ECG yields highly distinctive characteristics, suitable for various applications and diagnosis. This chapter focuses on cardiac arrhythmia classification. Cardiac arrhythmia is a heart disorder displaying an irregular heartbeat due to malfunction in the cells of the heart's electrical system. During cardiac arrhythmia, the heartbeat can have an irregular rhythm. Sometimes it is too fast – >90 beats/min – and this is called tachycardia; when the heartbeat is too slow – <60 beats/min – this is called bradycardia. Thus, there are many types of cardiac arrhythmia based on heart rate and site of origin. Some are frequently benign, although several may be a sign of significant heart disease, stroke, or surprising heart failure. At some stage in cardiac arrhythmia, the heart may not be capable of pumping enough blood to the body. Lack of blood flow can damage organs like the brain and heart. Thus, to enable appropriate survival measures, an accurate classification is required of cardiac arrhythmia that leads to heart rate variations. This chapter introduces the classification of six types of cardiac arrhythmias based on the adaptive neuro-fuzzy inference system (ANFIS).

This chapter is structured as follows. Section 1.2 gives a review of the literature. Section 1.3 describes system design and QRS complex detection and features the extraction method used in the classifier system. Section 1.4 describes system implementation using proposed methodology. Section 1.5 presents results and analysis and finally section 1.6 gives a discussion and conclusion.

DOI: 10.1201/9781003230540-1

1

1.2   REVIEW OF THE LITERATURE

Despite the ease of obtaining data, challenges remain for us to extract reliable information from biomedical signals. This can be a very demanding task for a computerized
automated system for several reasons. Among the noteworthy problems are
Biomedical signal contamination of the physiological tool and external noise as well
as unbalanced classes among biomedical signals. This causes the system performance
and accuracy to vary widely from patient to patient. The changing dynamics over
time and the morphological characteristics of the cardiac ECG signal show significant
variations between different patients in different physical and temporal conditions.

Linear discriminant analysis (LDA), artificial neural network (ANN), self-organ-
ization map (SOM), various clustering algorithms, radial basis function network
(RBFN), support vector machine (SVM), fuzzy systems, and other classifiers were
used in the examination survey. A few of these are briefly discussed below.

Mariano Llamedo and Juan P Matinez [1] used LDA to categorize five different
classes. The authors' main emphasis was on feature selection. A floating features
selection algorithm was used that performs better and generalizes the training and
validation sets. Overall, 93% accuracy was reached. The most commonly used classifier focused on ANNs with multiple-layer perceptron (MLP), probabilistic neural
network (PNN), and hybrid systems consisting of a combination of fuzzy clustering
with neural networks (NN), presented in papers [2–6].

For ECG classification, Inan Guler and Elif Derya Ubeyl [2] used combined NNs
with two levels of NN. The 19 statistical features obtained from the discrete wavelet
transform (DWT) decomposition coefficients were used to train the first stage of NN.
The second level of NN was trained to form the output of first-level NN. Various
algorithms were proposed based on the NNs with different structures and feature
vectors, and the results obtained have been shown to be state of the art. The methods
implemented were NN with adaptive activation function [3], recurrent NN (RNN)
with the Levenberg–Marquardt algorithm as optimization algorithm [4], combination
of continuous wavelet transform (CWT), principal component analysis and neural
network [5], and a combination of CWT, DWT, and dicrete cosine transform (DCT)
with MLP-back propagation (BP) [6].

To improve the efficiency of the classifier, several researchers suggested a hybrid
framework combining fuzzy systems and NN. Most studies [7–13] used a two-step
implementation, with fuzzy clustering as the feature vector's pre-processing and
MLP NN as the second-level classifier.

Yun-Chi Yeh et al. [10, 11] proposed a fuzzy logic model that included a qualitative
feature selection stage, fuzzy rule establishment, and classification stage. Nine morphological features were mapped to four quantitative features. The fuzzy rules with
triangular membership functions were defined using these four quantitative features.
To get the final output, a center of gravity defuzzification was used. Overall, 93.78%
accuracy was achieved. The authors also proposed a fuzzy C-mean classifier that used
the same features to identify the instances, with 93.57% accuracy.

The limitation of NN is that their performance depends on the activation function
used, the number of hidden layers and neurons, and the algorithm used for training
the NN.

SVM is "one of the most popular tools used for data classification and function approximation (Vapnik 1998a, 1998b)" due to its generalizability [14]. SVM can map input data into a high-dimensional feature space, transforming it into linearly separable data. Dimension reduction is essential for an efficient classifier, specifically for large datasets. Different approaches for data reduction were presented along with SVM [14].

The MLP NN was used to classify arrhythmia in most of the literature reviewed. The key disadvantage of MLP NN is that it is difficult to determine the number of hidden layers, the number of neurons to use for each hidden layer and the activation functions to use. Fixed-layer architecture is suggested as an improved version of MLP NN, in which the number of layers is fixed and the hidden and output layers can use different activation functions. The number of neurons in the hidden layer can also be calculated using a combination of different clustering methods, such as MLP NN, RBFN, and SVM.

Compared to MLP NN, SVM is more efficient in computational efficiency in terms of training time as well as testing time [15]. But the limitation of SVM is that the behavior of these structures for imbalanced classes is negative [16]. Also, the database-balancing techniques for the training phase have not been explored in depth. Overall, the learning of all these classifiers depends much more on how rich the dataset is used for training of the networks.

A notable contribution of neuro-fuzzy and subtractive clustering (SC) is the exposition of ANFIS, a system developed by Jang and which has found numerous applications in a variety of fields [17, 18]. ANFIS and its variants and relatives in the realms of neural, neuro-fuzzy, and reinforcement learning systems represent a direction of basic importance in the classification of ECG arrhythmias [19]. Hence this chapter proposes the implementation of classifiers based on the neuro-fuzzy approaches that incorporate the imprecise decision-making capability of fuzzy inference systems and the pattern recognition and adaptation properties of NN systems.

1.3 SYSTEM DESIGN

1.3.1 METHODOLOGY

Figure 1.1 shows a block schematic of the proposed classifier system from raw input signal processing to R-peak detection, features extraction and classification of the signal.

1.3.2 DATABASE USED

The database used for experimental analysis and classification was obtained from the Massachusetts Institute of Technology–Beth Israel Hospital's (MIT-BIH) arrhythmia database. It consists of 48 annotated ECG records from 47 subjects. Each record lasts 30 minutes with a sampling frequency of 360 Hz. All the records used were obtained from a modified limb lead II (MLII). Normal QRS complexes are usually prominent in the lead II signal [20].

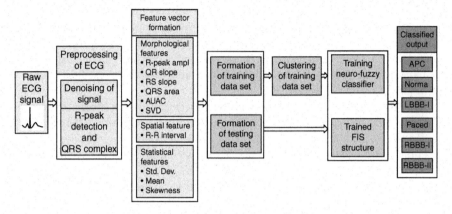

FIGURE 1.1   Block schematic of the system.

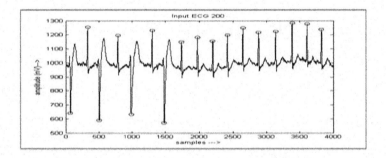

FIGURE 1.2   R-peak detection of record no. 200.

1.3.3   R-PEAK DETECTION AND QRS COMPLEX EXTRACTION

R-peak detection of an ECG signal is one of the most common and critical parts of the proposed method. The QRS complex was extracted from the raw ECG input once the R-peak had been correctly identified. The authors proposed a novel method for R-peak detection using curvelet transform (CT) and curvelet energy in reference [21]. Figure 1.2 shows the R-peak detected using CT. Detected peaks were validated by comparison with the annotation given for each record in the MIT-BIH arrhythmia database.

The entire QRS complex was extracted by considering 90 samples and 89 samples on either side of the located R-peak, known as a cardiac event or a heartbeat, as shown in Figure 1.3. Then Q and S peaks were located from the QRS complex, as shown in Figure 1.4, and these Q-, R-, and S-peak locations were used to obtain the features of the cardiac events [22].

1.3.4   FEATURE EXTRACTION

The QRS complex was used to derive the features used for training and evaluating the proposed classifier. Six morphological features, one spatial feature, and three statistical features were derived from this QRS complex. The morphological features

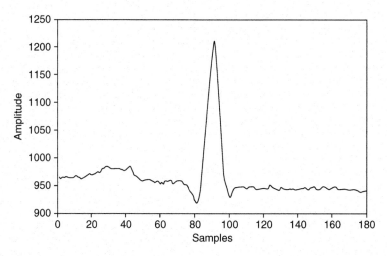

FIGURE 1.3 QRS complex extraction.

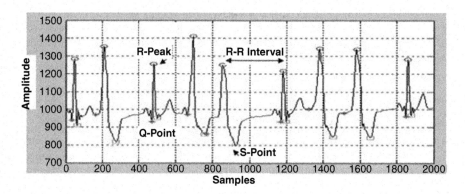

FIGURE 1.4 Q-R-S-peak location and R-R interval [22].

obtained included R-peak amplitude, area under auto-correlation, singular value decomposition, QRS area, QR slope, and RS slope. The spatial feature obtained was R-R duration and the statistical features were mean of QRS complex, standard deviation of QRS complex, and skewness of QRS complex.

1.4 SYSTEM IMPLEMENTATION

Figure 1.5a is the flow diagram for modeling of ANFIS using clustering algorithms and a hybrid learning algorithm. Figure 1.5b shows the cardiac arrhythmia classified by the ANFIS classifier.

In this chapter, two clustering methods are discussed to optimize the input features vector and to decide the initial premises parameters for Gaussian membership function which gives the initial FIS to the ANFIS. This FIS was fine-tuned (optimized

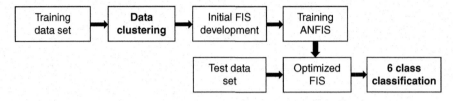

FIGURE 1.5A Modeling of adaptive neuro-fuzzy inference system (ANFIS).

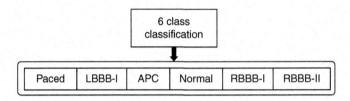

FIGURE 1.5B Cardiac arrhythmias classified by adaptive neuro-fuzzy inference system.

for required root mean square error (RMSE)) using PSO. The least-squares estimation (LSE) algorithm was used for tuning the linear consequence parameters. Once the ANFIS was trained it generated a final FIS that was used for testing the unseen dataset.

1.4.1 CLUSTERING METHODS

1.4.1.1 Subtractive Clustering

In the subtractive clustering process, every data point is presumed to be a potential cluster center at first. Every data point's Pi potential is determined in relation to the position of other data points (equation 1.1) [23].

$$P_i = \sum_{j=1}^{n} e^{-\propto \left\| x^i - x^j \right\|^2} \tag{1.1}$$

where $\propto = \dfrac{\gamma}{r_a}$, P_i is the ith cluster center (potential value of ith data), α is the weight between i^{th} data to j^{th} data, χ is the data point, γ is the variable set commonly to 4, and r_a is cluster radius.

A measure of the likelihood that each data point can identify the cluster center was obtained based on the potential density of surrounding data points [24]. The subtractive clustering algorithm was as follows:

1. The initial cluster center is chosen as a data point with a high potential value.
2. Then, by removing all data points from the first cluster center, the next data cluster and its center position are determined.
3. Repeat step 2 until all of the data points in the dataset had been covered.

Thus, predefined r_a defined the area of influence of every cluster center and these two parameters were used as initial premises parameters of FIS.

1.4.1.2 Fuzzy C-Means Clustering (FCM)

The most common parameter used to measure the similarities between objects for subtractive clustering is Euclidean distance. Subtractive clustering is the hard clustering technique that strictly classifies objects into different clusters. The major drawback of the hard clustering technique is loss of information for objects with less similarity. FCM permits one piece of fact to belong to two or more clusters. Fuzzy cluster analysis uses membership value to classify objects in a range from 0 to 1. The FCM algorithm was originally developed by Dunn in 1973 [25] and improved by Bezdek in 1981 [26].

The idea behind the FCM algorithm wass to partition the N-sampled training vectors into C clusters by minimizing the objective function, represented by equation (1.2) [27, 28].

$$ J(X;u,v) = \sum_{i=1}^{N} \sum_{j=1}^{C} \left(u_{i,j} \right)^{k} \left(\left\| X_i - v_j \right\| \right)^{2} \tag{1.2} $$

where k ($1 < k < \infty$) is the weighting exponent which determines the fuzziness of the cluster, $u_{i,j} u_{i,j}$ is the degree of membership of X_i in X_i the jth cluster v_j and should satisfy the conditions in equation (1.2), X_i is the ith d-dimensional feature vector, v_j is the d-dimensional center of the cluster and is calculated using equation (1.4), and $\|*\|$ is the Euclidian distance norm expressing the similarity between the measured data and the center with the range of various parameters as given in equation (1.3).

$$ 0 \leq u_{i,j} \leq 1, \quad 1 \leq i \leq N, \quad 1 \leq j \leq C \quad and \quad \sum_{j=1}^{C} \left(u_{i,j} \right) = 1, \quad 1 \leq kN \tag{1.3} $$

$$ v_j = \frac{\sum_{i=1}^{N} \left(u_{i,j} \right)^{k} \left(X_i \right)}{\sum_{i=1}^{N} \left(u_{i,j} \right)^{k}}; 1 \leq j \leq C \tag{1.4} $$

The algorithm of FCM is implemented as follows:

1. Initialize the matrix U = [$u_{i,j}$], U (0).
2. Calculate the center vectors v_j using equation 1.4.
3. Update the fuzzy membership based on the distance between the cluster center and each point in the matrix using equation 1.5.

$$ u_{i,j} = \frac{1}{\sum_{k=1}^{C} \left[\frac{\left\| X_i - v_j \right\|}{\left\| X_i - v_k \right\|} \right]^{\frac{2}{m-1}}} \tag{1.5} $$

4. Stop if $\left| u_{ij;k} - u_{ij;k+1} \right| < Threshold$; otherwise return to step 2.

Here the cluster centroid is obtained as the mean of all points, weighted through the degree of belonging to the cluster. The degree of belonging in a certain cluster is allied to the inverse of the distance to the cluster. FCM iteratively shifts the cluster centers to the "right" location within a dataset by updating the cluster centers and membership grades for each data point. Performance of the FCM algorithm depends on initial centroids.

1.4.2 BASIC ANFIS ARCHITECTURE

The ANFIS architecture is identical to a multilayer NN that has three hidden layers. The hidden layers of ANFIS represent membership functions and fuzzy rules. The first-order Sugeno model best represents the ANFIS architecture. The fuzzy rules used for ANFIS are represented using the first-order Sugeno model, as follows:

Rule 1: If ($x1$ is MA1) and ($x2$ is MB1) then (fn1 = p1 × 1 + q1 × 2 + c1).
Rule 2: If ($x1$ is MA2) and ($x2$ is MB2) then (fn2 = p2 × 1 + q2 × 2 + c2)

where $x1$ and $x2$ are the inputs to node i and MAi and MBi are the linguistic labels associated with this node. Figure 1.6 shows the ANFIS architecture that implements the fuzzy rules specified above. In this structure circles represent fixed nodes, whereas adaptive nodes are indicated by squares.

The input of layer 1 is converted into the fuzzy membership grade given by equation 1.6:

$$OP_{1,i} = MF_{Ai}(x1), \quad for\ i = 1,2\ or$$

$$OP_{1,i} = MF_{Bi-2}(x2), \quad for\ i = 3,4 \tag{1.6}$$

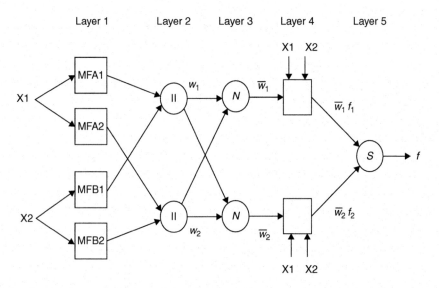

FIGURE 1.6 Adaptive neuro-fuzzy inference system architecture [26].

where $MF_{Ai}(x1)$, $MF_{Bi-2}(x2)$ can adapt to any fuzzy membership function and is used as shown in equation 1.7:

$$MF_A(x) = ex\left\{-\left(\frac{(x-c)^2}{2\sigma^2}\right)\right\} \tag{1.7}$$

where c and σ are nonlinear premises parameters of the membership function that represents the center and width of the Gaussian membership function. c is obtained from one of the clustering methods and σ is obtained by calculating the standard deviation of the data points lying in that cluster around the cluster center.

Nodes in the second layer are represented by π. The outputs of this layer are the product of all incoming membership functions, represented as set out in equation (1.8):

$$OP_{2,i} = W_i = MF_{Ai}(x1).MF_{Bi-2}(x2), \quad i = 1,2 \tag{1.8}$$

$OP_{2,i}$ gives the firing strengths of the rules.

Third-layer nodes are represented by N and indicate normalization of the firing strengths producing the output (equation 1.9).

$$OP_{3,i} = \overline{W_i} = \frac{W_i}{W_1 + W_2}, \quad i = 1,2 \tag{1.9}$$

The output of each node in the fourth layer is obtained by multiplying the output of layer 3 and a first-order polynomial for the Sugeno model, as given in equation (1.10).

$$OP_{4,i} = \overline{W_i}f_i = \overline{W_i}\left(p_i x1 + q_i x2 + c_i\right) \tag{1.10}$$

where $\{p_i, q_i, c_i\}$ represents the linear consequent parameters.

The fifth layer computes the overall output by adding all incoming signals given by equation (1.11).

$$OP_5 = \sum_i \overline{W_i}f_i = \frac{\sum_i W_i f_i}{\sum_i W_i} \tag{1.11}$$

1.4.3 ANFIS TRAINING

ANFIS training/learning means to tune the premises parameters and the consequent parameters of the network to obtain the desired output specified by the training data. Once trained the ANFIS has the capability of inference for the new input sets. Jang developed a hybrid learning algorithm for fast ANFIS parameters training [29]. It consists of two passes: (1) forward pass; and (2) backward pass.

The premises parameters are unchanged during the forward pass and consequent parameters are modified according to the LSE algorithm given by equation 1.12:

$$\hat{\theta} = \left(A^T W A\right)^{-1} A^T W y \qquad (1.12)$$

where $\hat{\theta}$ is estimation matrix of consequence parameter, W is weight vector, A is input feature matrix, and y is expected output.

The consequence parameters are fixed during the backward pass, and the premises parameters are tuned using the proposed PSO technique. These two passes are iterated for a set number of epochs or until the error requirements are met.

1.4.3.1 Particle Swarm Optimization Algorithm

The key benefit of using PSO as a training algorithm is that it is derivative-free. As a result, the algorithm's computational complexity is lower than that of derivative-based approaches.

Unlike other optimization techniques it is less dependent on a set of initial points. In general, there are three major algorithms available in the PSO. The first is the individual best. Each individual compares their position to their own best position, p_{best}. These algorithms do not use knowledge from other particles. The second version is the global best. The location of the best particle from the entire swarm is included in the social information used to direct particle movement. Furthermore, each particle considers its previous experiences in terms of its own best solution.

1.4.3.1.1 Global-Best PSO

In the entire swarm the best-fit particle impacts the location of each particle in this process. Each individual particle $i \in [1,\ldots,n]$ where $n > 1$ in the search space has a current velocity v_i, a current position x_i, and a personal-best position $P_{best,i}$. When considering a minimization problem, the position in the search space where particle i has the least value as defined by the goal function is the personal-best position. Furthermore, the position with the lowest value of all is referred to as the global-best position, which is denoted as [30]. Equations 1.13 and 1.14 are used to update the personal-best and global-best places, respectively.

$$P_{best,i}^{t+1} = \{ P_{best,i}^t \quad if \ f(x_i^{t+1} > P_{best,i}^t) x_i^{t+1} \quad if \ f\left(x_i^{t+1} \leq P_{best,i}^t\right) \qquad (1.13)$$

Equation 1.14 is used to calculate the global-best position g_{best} at time step t.

$$g_{best} = min\left\{ P_{best,i}^t \right\} \qquad (1.14)$$

where $i \in [1,\ldots,n]$ and $n > 1$.

The velocity of a particle is calculated using equation 1.15 for the global-best PSO method.

$$v_{ij}^{t+1} = v_{ij}^t + c_1 r_{1j}^t \left[P_{best,i}^t - x_{ij}^t \right] + c_2 r_{2j}^t \left[g_{best} - x_{ij}^t \right] \qquad (1.15)$$

where $v_{ij}^t v_{ij}^t$ is the particle's current velocity vector in j dimension, $x_{ij}^t x_{ij}^t x_{ij}^t$ is the particle's current position vector in j dimension, $P_{best,i}^t$ is the particle's current personal-best position in j dimension, g_{best} is the particle's global-best position in j dimension, positive acceleration constants c_1 and c_2 are employed to balance the contributions of the cognitive and social components, respectively, and r_{1j}^t and r_{1j}^t are random values from the uniform distribution U (0,1).

1.4.3.1.2 Local-Best PSO (lbest)

The circle neighborhood structure is reflected in the local best (l_{best}). Particles are influenced by their surroundings as well as their own prior experiences. Equation 1.16 is used to calculate the velocity of the particle i where g_{best}, the global-best value, is substituted by l_{best}.

$$v_{ij}^{t+1} = v_{ij}^t + c_1 r_{1j}^t \left[P_{best,i}^t - x_{ij}^t \right] + c_2 r_{2j}^t \left[l_{best} - x_{ij}^t \right] \qquad (1.16)$$

where l_{best} is the best position of a particle in the neighborhood of the particle i found from initialization through time t.

l_{best} is slower in convergence than g_{best}. l_{best} results in a much better solution [31] and searches a larger part of the search space.

Shi and Eberhart [30] have modified equations 1.15 and 1.16 by introducing an inertia weight w as in equation 1.17:

$$v_{ij}^{t+1} = w v_{ij}^t + c_1 r_{1j}^t \left[P_{best,i}^t - x_{ij}^t \right] + c_2 r_{2j}^t \left[g_{best} - x_{ij}^t \right] \qquad (1.17)$$

The inertia weight w is responsible for balancing global and local searches, and its value can change in the optimization process. A high inertia weight favors a global search, while a low value favors a local search [32].

The algorithm to implement modified PSO is as follows:

1. At $t = 0$, initialize the swarm of particles $P(t)$ such that for each particle $Pi \in P(t)$ has a random position within the hyperspace.
2. Evaluate each particle's performance based on its current position $\vec{x}_i(t)$.
3. Compare each individual's performance to their best performance.

$$if \left(F\left(\vec{x}_i(t) \right) < P_{best} \right) then\, P_{best,i} = F\left(\vec{x}_i(t) \right) else\, P_{best,i} = \vec{x}_i(t)$$

4. Compare each individual's performance to their global-best particle.

$$if \left(F\left(\vec{x}_i(t) \right) < g_{best} \right) then\, Pg_{best} = F\left(\vec{x}_i(t) \right) else\, g_{best} = \vec{x}_i(t)$$

5. Using equation 1.17, change each particle's velocity vector.
6. Update the position of each particle using equation 1.18.

$$x_{ij}(t+1) = x_{ij}(t) + v_{ij}(t+1) \qquad (1.18)$$

7. Repeat steps 2–6 until convergence/number of iterations specified.

1.4.3.2 Experimental Parameter Setting of Clustering and Training Algorithms

As discussed in the algorithm the initial values of parameters to implement the required algorithm are set as follows.

Initial Parameters Set for PSO and FCM for ANFIS Classifiers
Maximum of iterations (MaxIt = 5000), inertia weight (w = 0.01), inertia weight damping ratio (wdamp = 0.9), population size/swarm wize (nPop = 30), Global learning coefficient (c2 = 3), cost function used is RMSE.

FCM Parameters
Exponent for the partition matrix U (fcm_U = 5), fcm_MaxIter = 5000, fcm_MinImp = 1e-5.

Initial parameters set for PSO and SC for ANFIS classifiers
Maximum of iterations (MaxIt=5000), inertia weight (w = 1), inertia weight damping ratio (wdamp = 0.9), population size/swarm zize (nPop = 30), personal learning coefficient (c1 = 0.1), global learning coefficient (c2 = 3), cost function used is RMSE.

Subtractive Clustering: Radii = 0.5.

1.5   RESULTS AND ANALYSIS

The training and testing datasets are used to train, evaluate, and analyze the proposed models. The 600-exemplar training dataset is generated by randomly selecting 100 samples of each type of cardiac arrhythmia, including normal beat type, for each type. Similarly, the testing dataset is generated excluding the feature vectors used for the training dataset. Sensitivity, specificity, positive predictivity (PP), false prediction ratio (FPR), and classification rate (CR) are the five performance parameters used to test and analyze the classifier [23].

The results obtained for SC-ANFIS-PSO and FCM-ANFIS-PSO classifiers are presented below in the form of confusion matrix and performance parameters. Finally, the average values of all performance parameters are summarized for both the models.

1.5.1   RESULTS OF SIX-CLASS SC-ANFIS-PSO CLASSIFIER

Tables 1.1 and 1.2 show the confusion matrix and performance parameters for the SC-ANFIS-PSO model respectively.

1.5.2   RESULTS OF SIX-CLASS FCM-ANFIS-PSO CLASSIFIER

Figure 1.7 shows various error plots while training the classifier. It shows the target output vs. actual output, error plot, and distribution of error with various statistical parameters.

TABLE 1.1
Confusion matrix

	P	L-I	A	N	R-I	R-II
P	94	5	0	0	0	0
L-I	0	95	3	1	0	0
A	0	3	92	3	1	0
N	1	2	13	78	5	0
R-I	0	1	0	2	95	0
R-II	0	0	0	5	1	91

TABLE 1.2
Class-wise performance parameters

Classes	Sensitivity	Specificity	PP	FPR	CR
P	0.95	1.00	0.99	0.00	0.99
L-I	0.96	0.98	0.90	0.02	0.97
A	0.93	0.97	0.85	0.03	0.96
N	0.79	0.98	0.88	0.02	0.95
R-I	0.97	0.99	0.93	0.01	0.98
R-II	0.94	1.00	1.00	0.00	0.99

PP, positive predictivity; FPR, false prediction ratio; CR, classification rate.

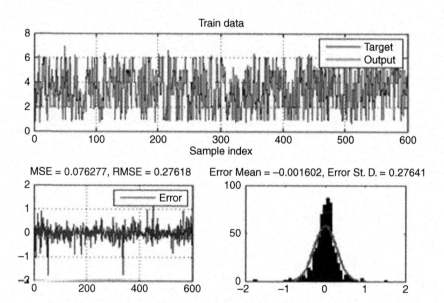

FIGURE 1.7 Target output vs. actual output, error plot, and error distribution.

TABLE 1.3
Confusion matrix

	P	L-I	A	N	R-I	R-II
P	99	1	0	0	0	0
L-I	1	95	3	1	0	0
A	0	3	97	0	0	0
N	0	0	20	57	1	0
R-I	0	0	0	12	87	0
R-II	0	0	0	3	8	89

TABLE 1.4
Class-wise performance parameters

Classes	Sensitivity	Specificity	PP	FPR	CR
P	0.99	1.00	0.99	0.00	1.00
L-I	0.95	0.99	0.96	0.01	0.98
A	0.97	0.95	0.81	0.05	0.95
N	0.73	0.97	0.78	0.03	0.94
R-I	0.88	0.98	0.91	0.02	0.96
R-II	0.89	1.00	1.00	0.00	0.98

PP, positive predictivity; FPR, false prediction ratio; CR, classification rate.

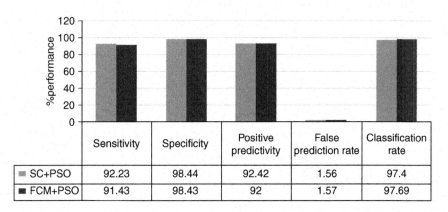

	Sensitivity	Specificity	Positive predictivity	False prediction rate	Classification rate
SC+PSO	92.23	98.44	92.42	1.56	97.4
FCM+PSO	91.43	98.43	92	1.57	97.69

FIGURE 1.8 Comparison of performance parameters of adaptive neuro-fuzzy inference system classifier with particle swarm optimization (PSO).

Tables 1.3 and 1.4 show the confusion matrix and performance parameters for the trained FCM-ANFIS-PSO model respectively

Figure 1.8 shows the comparison of average performance parameters of two proposed ANFIS-PSO classifier models. The training time for both models has

TABLE 1.5
Comparison of proposed methods with survey methods

Sr. No.	Literature	Classifier types	No. of classes	Feature vector size	Classification accuracy/ sensitivity	Training data size
1	This work	SC + ANFIS + PSO	6	10	97.4%	600
2	This work	FCM + ANFIS + PSO	6	10	97.69%	600
3	Edoardo Pasoli and Farid Melgani (2010) [32]	SVM	06	12 samples	95%	2000
4	Mariano Llamedo and Juan P Matinez (2011) [1]	Linear discriminant classifier	05	8 Samples	93%	@50000
5	Yun-Chi Yeh et al. (2009) [10]	Fuzzy Logic Method	05	9	93.78%	-

SC, subtractive clustering; ANFIS, adaptive neuro-fuzzy inference system; PSO, particle swarm optimization; FCM, fuzzy C-means clustering; SVM, support vector machine.

been observed: 792 seconds for the SC-ANFIS-PSO model and 849 seconds for the FCM-ANFIS-PSO model. Analysis of the proposed method shows that SC + PSO model results are better in terms of training time whereas performance parameters as compared to the FCM + PSO model are almost the same.

Table 1.5 is a comparison of the proposed method with the methods described in the literature. It can be clearly observed from this comparison that the proposed method yields good classification accuracy with smaller training data size as well as feature vector.

1.6 CONCLUSION

Analysis of the results and comparison of the two proposed models shows that both models perform equally well and are nearly identical in terms of performance parameters. The neuro-fuzzy classifier integrates the benefits of fuzzy logics in human knowledge expression with a computational self-adaptation learning process. Comparison with the survey methods also shows state-of-the-art performance with the proposed method.

The beauty of the proposed system is that the fuzzy rules do not need to be physically set. The ANFIS model will automatically set the rules, which will be fine-tuned using a hybrid learning algorithm. The computation complexity is further reduced by PSO which is a non-derivative-based back propagation training algorithm.

All these parameters lead to better performance parameters with average classification rate of 97.4% and 97.69% for SC-ANFIS-PSO classifier and FCM-ANFIS-PSO classifier, respectively. As a consequence of the observations and analysis, the proposed approach can be used to automatically classify cardiac arrhythmias with higher classification accuracy and lower false prediction rates.

1.6.1 FUTURE SCOPE

The proposed neuro-fuzzy models are multiple-input and single-output-type networks where only one output node is used to represent all the classes. Each class is labeled as a number ranging from 1 to N, where N is the total number of classes to be classified. Further modifications in the ANFIS structure may be made to obtain the N-different outputs corresponding to each class. The number of cardiac arrhythmias that can be classified can also be increased to more than six.

REFERENCES

[1] Mariano Llamedo and Juan P Matinez, "Heartbeat classification using feature selection driven by database generalization criteria," *IEEE Transactions on Biomedical Engineering* 58 (2011), 616–625.

[2] Inan Guler and Elif Derya Ubeyl, "ECG beat classifier designed by combined neural network model," *Pattern Recognition* 38 (2005), 199–208.

[3] Yuksel Ozbay and Gulay Tezel, "A new method for classification of ECG arrhythmias using neural network with adaptive activation function," *Digital Signal Processing* 20 (2010), 1040–1049.

[4] Elif Derya Ubeyli, "Recurrent neural networks employing Lyapunov exponents for analysis of ECG signals," *Expert Systems with Applications* 37 (2010), 1192–1199.

[5] Parham Ghorbanian, Ali Ghaffari, Ali Jalali and C Nataraj, "Heart arrhythmia detection using continuous wavelet transform and principal component analysis with neural network classifier," *IEEE Journal of Computing in Cardiology* (2010), 669–672.

[6] Hamid Khorrami and Majid Moavenian, "A comparative study of DWT, CWT and DCT transformations in ECG arrhythmias classification," *Expert Systems with Applications* 37 (2010), 5751–5757.

[7] Stanislaw Osowski and Tran Hoai Linh, "ECG beat recognition using fuzzy hybrid neural network," *IEEE Transactions on Biomedical Engineering* 48 (2001), 1265–1271.

[8] Mehmet Engin, "ECG beat classification using neuro-fuzzy network," *Pattern Classification Letters* 25 (2004), 1715–1722.

[9] Liang-Yu Shyu, Ying-Hsuan Wu and Weichih Hu, "Using wavelet transform and fuzzy neural network for VPC detection from the Holter ECG," *IEEE Transactions on Biomedical Engineering* 51 (2004), 1269–1273.

[10] Yun-Chi Yeh, Wen-June Wang and Che Wun Chiou, "Heartbeat case determination using fuzzy logic method on ECG signals," *International Journal on Fuzzy Systems* 11 (2009), 250–260.

[11] Yun-Chi Yeh, Wen-June Wang and Che Wun Chiou, "A novel cuzzy C-means method for classifying heartbeat cases from ECG signals," *Measurements* 43 (2010), 1542–1545.

[12] Rahime Ceylan, Yuksel Ozbay and Bekir Karlik, "A novel approach for classification of ECG arrhythmias: Type-2 fuzzy clustering neural network," *Expert Systems with Applications* 36 (2009), 6721–6726.

[13] Yuksel Ozbay, Rahime Ceylan and Bekir Karlik, "Integration of type-2 fuzzy clustering and wavelet transform in a neural network based ECG classifier," *Expert Systems with Applications* 38 (2011), 1004–1010.

[14] Nurettin Acır, "A support vector machine classifier algorithm based on a perturbation method and its application to ECG beat recognition systems," *Expert Systems with Applications* 31 (2006), 150–158.

[15] C. Cortus and V. Vapnik, "Support-vector networks," *Machine Leaming* 20 (1995), 273–297.

[16] li Zughrat, M. Mahfouf, Y.Y. Yang and S. Thornton, "Support vector machines for class imbalance rail data classification with bootstrapping-based over-Sampling and under-sampling," *IFAC Proceedings Volumes* 47 (2014), 8756–8761.

[17] J.S.R. Jang, "ANFIS: Adaptive-network-based fuzzy inference systems," *IEEE Transactions on Systems, Man and Cybernetics* 23 (1993), 665–685.

[18] Ahmet Yardimci, "Soft computing in medicine," *Applied Soft Computing* 9 (2009), 1029 –1043.

[19] M. Aliyari Shoorehdeli, M. Teshnehlab and A. K. Sedigh, "A novel training algorithm in ANFIS structure," *Proceedings of the 2006 IEEE American Control Conference Minneapolis,* Minnesota, USA, June 14–16, 2006.

[20] S. Tandale, A. S. Barhatte, R. Ghongade and M. Dale, "Arrhythmia classification using neuro fuzzy approach," *2017 3rd International Conference on Advances in Computing, Communication and Automation (Fall), Dehradun,* India, 2017, pp. 1–4.

[21] A. Barhatte, M. Dale and R. Ghongade, "Cardiac events detection using curvelet transform," *Sādhanā* 44 (2019), 47. doi: 10.1007/s12046-018-1046-0

[22] A. S. Barhatte, R. Ghongade and A. S. Thakare, "QRS complex detection and arrhythmia classification using SVM," *2015 Communication, Control and Intelligent Systems (CCIS)*, Mathura, India, 2015, pp. 239–243. doi: 10.1109/CCIntelS.2015.7437915.

[23] A. Barhatte and R. Ghongade, "A multiclass cardiac events classifier using clustering and modified adaptive neuro-fuzzy inference system," *2015 IEEE International Conference on Research in Computational Intelligence and Communication Networks (ICRCICN)*, Kolkata, India, 2015, pp. 90–95. doi: 10.1109/ICRCICN.2015.7434216.

[24] Agus Priyono, Muhammad Ridwan, Ahmad Jais Alias, Riza Atiq O. K. Rahamt, Azim Hassan and Mohd. Alauddin Mohd. Ali, "Generation of fuzzy rules with subtractive clustering," *Jurnal Teknologi*, 43 (2005), 143–153.

[25] J. C. Dunn, "A fuzzy relative of the ISODATA process and its use in detecting compact well-separated clusters," *Journal of Cybernetics* 3 (1973), 32–57.

[26] J. C. Bezdek, *Pattern Recognition with Fuzzy Objective Function Algorithms,* Plenum Press, New York, 1981.

[27] Yun-Chi Yeh, Wen-June Wang and Che Wun Chiou, "A novel fuzzy C-means method for classifying heartbeat cases from ECG signals," *Measurement* 43 (2010), 1542–1555.

[28] W. C. Chen and M. S. Wang, "A fuzzy C-means clustering-based fragile watermarking scheme for image authentication," *Expert System with Application* 36 (2009), 1300–1307.

[29] E. D. Ubeyli and I. Guler, "Adaptive neuro-fuzzy inference system to compute quasi-TEM characteristic parameters of microshield lines with practical cavity sidewall profiles," *Neurocomputing* 70 (2006), 196–204. http://basicknowledge101.com/subjects/physicalhealth.html

[30] Y. Shi and R. C. Ebenhart, "Empirical study of particle swarm optimization," *IEEE Congress on Evolutionary Computation* 3 (1999), 1945–1950.

[31] M. Mahfouf, M. Y. Chen and D. A. Linkens, "Adaptive weighted particle swarm optimisation for multi-objective optimal design of alloy steels," in: Yao X. et al. (eds.) *Parallel Problem Solving from Nature – PPSN VIII. PPSN 2004. Lecture Notes in Computer Science*, vol. 3242, Springer, Berlin, 2004.

[32] Edoardo Pasoli and Farid Melgani, "Active learning methods for electrocardiographic signal classification," *IEEE Transactions on Information Technology in Biomedicine* 14 (2010), 1405–1415.

2 Two-Stage Deep Learning Architecture for Chest X-Ray-Based COVID-19 Prediction

Mainak Chakraborty
Defence Institute of Advanced Technology, Girinagar,
Pune, India

Sunita Vikrant Dhavale
Defence Institute of Advanced Technology, Girinagar,
Pune, India

Jitendra Ingole
Smt. Kashibai Navale Medical College and General Hospital,
Narhe, Pune, India

2.1 INTRODUCTION

The novel coronavirus (COVID-19) disease outbreak began in December 2019 in Wuhan, China, and spread rapidly amongst citizens living in other countries [39,16,28,30]. Around 220 countries, areas, or territories were affected by novel coronavirus up to 8 December 2020, of which around 128 countries, areas, or territories had begun community transmission [38]. Globally, 66,729,375 confirmed COVID-19 cases, including 1,535,982 deaths, were reported to the World Health Organization (WHO) as of 8 December 2020 [38]. COVID-19 is caused by the SARS-CoV-2 virus, which spreads among individuals, especially when they are in direct or close contact with an infected person [37]. In small liquid particles, the virus can transmit from the mouth or nose of an infected person as they sneeze, sing, cough, speak, or breathe heavily [37]. The more common signs of COVID-19 are fatigue, dry cough, and fever; other less common symptoms include dizziness, nausea, headache, nasal congestion, diarrhea, joint pain, skin rash, sore throat, loss of taste or smell, and conjunctivitis [36]. To date, real-time reverse transcription polymerase chain reaction (RT-PCR) is considered the gold-standard method of coronavirus diagnosis [8,31] However, rapid antigen tests are also used to diagnose COVID-19, where samples are obtained with a swab from the nose and/or throat [36]. Rapid antigen testing is quicker and cheaper than RT-PCR [36]. Computed tomography (CT) images of

DOI: 10.1201/9781003230540-2

patients are also used to assess the severity of COVID-19 lung infection and for prognostication. The hallmarks of COVID-19 patients present in CT images are ground-glass opacity (91% vs. 68%, $p < 0.001$), peripheral distribution (80% vs. 57%, $p < 0.001$), vascular thickening (59% vs. 22%, $p < 0.001$), and fine reticular opacity (56% vs. 22%, $p < 0.001$) as compared with non COVID-19 patients [3].

However, there are problems with the current COVID-19 diagnostic techniques as follows:

- RT-PCR testing is costly and time consuming [1].
- RT-PCR test labs cannot meet the testing demand as the number of cases has been increasing day by day [1].
- The RT-PCR test has a specificity rate of almost 100%, but there is a 30–35% risk that false negatives will occur [1].
- Rapid antigen testing is generally less accurate than RT-PCR [36], and sensitivity of rapid antigen testing is around 50% [1].
- The sanctity of rapid antigen tests is around 1 hour from the time the sample is collected and analyzed since it cannot be performed at any collection center [1].
- CT scan devices are expensive and need a significant level of handling experience.
- In several medical centers in non-urban areas, high-quality CT scan equipment may not be available.
- Portability is another drawback of CT devices; thus, it is not a suitable imaging device for rapid testing during a pandemic situation.

To overcome the above limitations, recently, a chest X-ray (CXR)-based coronavirus diagnosis method was proposed by several researchers worldwide and it was proven that an X-ray-based diagnosis technique is a reliable tool to promote rapid testing [34,21,14,22,19,2]. X-ray imaging is cost-effective, fast, and easily accessible, so medical centers can use chest X-ray-based COVID-19 diagnosis in non-urban areas to isolate suspected patients immediately before confirmation by RT-PCR testing.

In recent years, deep learning-based disease prediction systems have become very popular and achieve promising results in various domains such as detection of thorax disease [35,24], arrhythmia [23], breast cancer [4], diabetic retinopathy [13], and pulmonary tuberculosis [18]. The deep learning technique can reduce the workload, cost, and diagnosis time of radiologists and doctors. Researchers are also trying to remove regions that are not relevant to the lungs or other regions to improve the accuracy of predictions [12,21]. Such a task is involved in the segmentation of the right and left lung contours from conventional CXR images [12,21].

This study aims to explore the deep convolutional neural networks (DCNNs) further and evaluate their diagnostic feasibility for coronavirus. Unfortunately, it is hard to collect many good-quality COVID-19 X-ray samples in a pandemic situation to train, validate, and test a DCNN module. Therefore, this article's primary objective is to achieve better sensitivity than RT-PCR with a limited and unbalanced training dataset for the diagnosis of coronavirus. This work is an extension of our preliminary journal paper [5]. Based on the previous version, this chapter proposes a two-stage

DCNN to improve the sensitivity of coronavirus disease detection. The motivation of the study is that if we can extract the lung contours from the CXR samples precisely, this reduces the diagnosis time and cost and allows the X-ray samples to be examined more easily. It also helps to efficiently train the deep neural network (DNN) model, even with the limited number of training samples. The results shows that the suggested model outperforms other state-of-the-art work and achieves a sensitivity of 99% for coronavirus disease and test accuracy of 98%, which was an acceptable performance as a diagnosis method compared to RT-PCT, which has a sensitivity of about 70% [1]. Ultimately, this low-cost, intelligent two-stage DCNN offers a detailed understanding that can contribute to the improvement of medical diagnostic decisions being made by medical professionals.

The remainder of this chapter is organized as follows. We discuss work related to the study and existing problems in Section 2.2. The architecture of the proposed DCNN and its implementation details are addressed in Section 2.3. In Section 2.4, experimental setup, dataset, results, and analysis, comparisons with other state-of-the-art works and performance on Indian patients' datasets are mentioned. Finally, Section 2.5 summarizes the conclusion and future outcomes of our study.

2.2 PROBLEMS AND RELATED WORK

Recent progress in computer vision research, particularly medical image analysis and disease diagnosis tasks [35,24,4], identification of human activity [27,7], smart city management [33,6], and the convolutional neural network (CNN) using a wide range of open-source datasets, has achieved promising results without the need for human intervention and manual feature extraction. Nowadays, many screening techniques are used to identify suspected lung disease lesions. X-rays are mostly considered an outdated form of medical imaging; however, using automated methods and deep learning reawakens X-ray's relevance for medical diagnostic imaging [12]. With the developments in DCNN and graphics processing unit (GPU) computing research, identifying more types of cardiothoracic lesions on X-ray scans became easier [12]. The availability of many open-source radiology images encourages researchers to apply their deep learning algorithms to analyze those images and find several disease-related hallmarks. To promote research in artificial intelligence (AI)-based diagnosis of tuberculosis using CXR samples, Montgomery County and Shenzhen Hospital X-ray datasets [17] are made available by the U.S. National Library of Medicine. On the other hand, the JSRT dataset [26] has been made available to the research community by the Japanese Radiological Society in association with the Japanese Society of Radiological Technology. ChestX-ray14 [35] is currently the largest open-access database of CXRs, featuring over 112,120 frontal X-ray images of 14 distinct lung diseases. Rajpurkar et al. [24] trained CheXNet, a 121-layer CNN using the aforementioned ChestX-ray14 dataset, to automatically diagnose pneumonia from X-ray samples to a degree beyond that of experienced radiologists. Gaál et al. [11] introduced U-Net-based adversarial CNN to extract lung segments from CXR samples. They trained and validated their proposed model using the JSRT dataset [26] and achieved 97.5% dice score coefficient. Gordienko et al. [12] utilized U-Net-based CNN to extract lung segments and exclude bone shadow from X-ray samples to analyze suspicious lesions and nodules present in the X-ray images of lung cancer patients. They trained and validated their model using the JSRT dataset [26] and achieved promising results. They also reported that pre-processed CXR images without rib shadows and clavicles give better results than pre-processed segmented lung images. Stirenko et al. [29] utilized the U-Net network suggested by Gordienko et al. [12] to analyze tuberculosis disease from extracted lung segments. They used the Shenzhen Hospital CXR dataset [17] and the corresponding manually segmented lung masks prepared by them to train the model.

In computer-assisted medical image analysis, recently, several DNN-based approaches obtained promising results to assess coronavirus infected patients by profound X-ray image analysis. Wang, Lin et al. [34] suggested COVID-Net, a DCNN, and attained 92.6% classification accuracy in detecting normal, pneumonia, and COVID-19 patients. They trained their model using a COVIDx dataset collected from three distinct open-access databases [10,9,20]. The dataset comprised 5,538 pneumonia, 8,066 normal, and 183 COVID-19 samples. We found that the COVID-Net DCNN suffers from many false-negative results for the coronavirus class. To overcome the limitation of the availability of coronavirus patients' X-ray samples, Oh et al. [21] designed a patch-based DCNN to detect bacterial pneumonia, normal, tuberculosis, viral pneumonia, and coronavirus disease using X-ray samples. The suggested approach is motivated by a statistical study of the CXR images' possible imaging hallmarks. They utilized FC-DenseNet103 and ImageNet pre-trained ResNet-18 CNN as a backbone of their segmentation and classification network, respectively. Further, random patches were generated from the extracted lung segments obtained from their segmentation network. The generated patches were then fed into the classification network for disease prediction. Transfer learning-based coronavirus patient detection was proposed by Hemdan et al. [14]. They found that the DenseNet121 and VGG19 CNN's experimental results were better than other CNNs. Even so, we find that with only 25 normal and 25 COVID-19 X-ray images, these CNN models are trained. Ozturk et al. [22] introduced the DarkCovidNet network to detect normal, pneumonia, and coronavirus disease using frontal CXR samples. They designed a DCNN model consisting of 17 convolutional layers inspired by DarkNet architecture. The model trained with minimal COVID-19 images and attained 87.02% accuracy. The model likely misses the significant biomarkers of the normal and pneumonia categories because of under-sampling strategies. Mangal et al. [19] obtained 90.5% test accuracy for CXR-based COVID-19 screening by proposing the CovidAID model. However, for normal patients, CovidAID CNN's false-negative prediction rate is high. Apostolopoulos and Mpesiana [2] utilized transfer learning in order to diagnose COVID-19 patients using 504 normal, 224 COVID-19, and 700 pneumonia samples. They achieved 93.48% classification accuracy. Hira et al. [15] utilized nine pre-trained CNNs and fine-tuned the CNNs using minimal CXR images. They reported that Se-ResNeXt-50 CNN outperformed other CNNs by achieving 96.99% three-class classification accuracy. All of these studies have obtained promising results by using thoracic X-rays to diagnose COVID-19-infected people. However, most of the early work has the following problems:

- Models are trained using a limited number of non-COVID samples.
- To overcome the unbalanced dataset problem, researchers utilized the under-sampling technique or priority-wise class weight that indirectly loses some crucial imaging hallmarks of the majority class (non-COVID).
- There were high false-positive and/or false-negative rates.
- Coronavirus is a respiratory disease [3,34,21,14,22,19,2] caused by the SARS-CoV-2 virus [37]. Therefore, training a deep learning module with CXR without segmentation may decrease the sensitivity of COVID-19 diagnosis because examining non-pulmonary areas in CXR images is not relevant to a study of coronavirus.

So, motivated by this prior success and to overcome the issues mentioned above, this article proposed a two-stage DCNN to diagnose coronavirus diseases utilizing CXR images. To tackle the imbalanced dataset problem, we utilized the minority class over-sampling method suggested in our earlier work [5]. We trained our model with the maximum number of open-source non-COVID samples to make the model more generalized. This work's main intuition is to learn only the hallmarks of coronavirus disease present in the CXR samples and evaluate the feasibility of the proposed model for COVID-19 diagnosis. Further, the model performance was tested in Indian coronavirus patients.

2.3   PROPOSED APPROACH

This section presents the architecture of the proposed model and its implementation details. The proposed DCNN architecture is shown in Figure 2.1. This work is motivated by the idea that precisely extracting the region of interest (ROI) area from CXR images can minimize diagnostic time and simplify CXR image analysis. The implemented model is divided into two parts: (1) segmentation network; and (2) classification network. The segmentation network extracts the ROI area, i.e., lung contours from the CXR samples, as shown in Figure 2.1(a). The classification network shown in Figure 2.1(b) predicts the corresponding lung disease based on analysis of the clinically relevant intricate infection patterns found in the contours of the extracted lungs. Both of the networks are implemented using Python 3.7.7, TensorFlow-GPU v1.14.0, Keras 2.2.4 API, and OpenCV 4.1.1.

2.3.1   SEGMENTATION NETWORK

The main objective of the segmentation network is to precisely extract the lung segment from the CXR samples. As we know in cases of any lung disease, the infected body parts are only the lung contours so other body parts present in X-ray images are not relevant for this study.

The segmentation network uses U-Net [25] as its backbone. We trained and validated the U-Net model using 951 CXR samples and their corresponding lung segments collected from different open-source datasets [17,26,29,32]. Before feeding the collected images to the U-Net model, initial image preprocessing, like image resize = (256,256,1), rescale with a 1./255 factor, gray-scaling and data augmentation

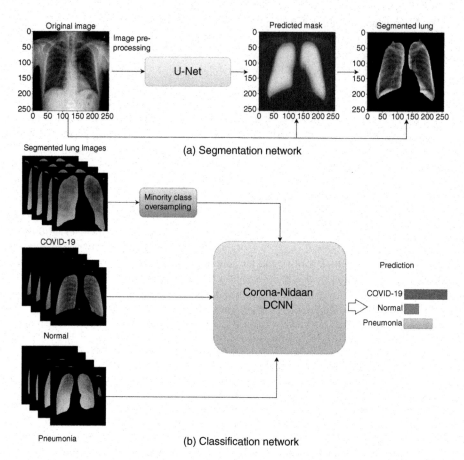

(a) Segmentation network

(b) Classification network

FIGURE 2.1 The proposed deep convolutional neural network (DCNN) architecture for COVID-19 diagnosis.

like rotation range = 10.0, zoom range = 0.2, height shift range = 0.2, horizontal flip = false, width shift range = 0.2, was applied to the raw CXR images. We used the normal kernel initializer, same padding and ReLU activation for all two-dimensional (2D) convolutional layers except the output layer. The output 2D convolutional layer is activated by sigmoid activation. The optimization algorithm and other hyperparameters of the segmentation networks are determined utilizing manual search techniques. We found that the model performed well with Adam optimizer, 1e^{-4} learning rate and four-batch size. To avoid overfitting we employed early stopping with the parameters patience = 20, min delta = 0.00001, and auto mode. During training, the binary cross-entropy loss function was used to minimize loss by optimizing the network.

2.3.2 Classification Network

The main goal of the classification network was to learn the hallmarks of respiratory infections from the segmented contour of the lungs and classify them according to the categories of lung disease. The classification network adopted the Corona-Nidaan [5] DCNN as its backbone. The proposed model first extracted the lung segments of patients with COVID-19, normal, and pneumonia through the segmentation network and then fed the extracted lung segments into the classification network. During data collection, we found that the number of COVID-19 samples was limited compared to pneumonia and normal CXRs. To address this problem we employed a minority class over-sampling approach and data augmentation techniques, as mentioned in our previous study [5]. The lightweight Corona-Nidaan architecture comprises 91 layers and was developed using five architectural design principles, and was inspired by MobileNetV2, InceptionResNetV2, and InceptionV3 architectures. To minimize the number of trainable parameters and computational cost, the model utilized depth-wise separable convolution layers. The model contains 4,021,974 parameters that are less than 1.0 MobileNet-224. The model adopted batch normalization after the convolutional layers to improve the generalization error and speed up training. The model consists of three I-blocks; each I-block learns multi-level feature representation from the input by applying different dimension filters simultaneously. Instead of a dense layer before the softmax layer, the model adopted a global average pooling layer to minimize the model's trainable parameters.

The segmented lung images were resized to (256,256,3) and rescaled with a 1./255 factor before training the model. A manual search technique was utilized to find the optimization algorithm and adjust the hyperparameters of the model. We found that with the Adam optimizer with an initial learning rate of 0.001 and eight batch sizes, the model performed well. The maximum epoch to train the model and monitor the validation accuracy was set to 300. The learning rate reduced by 0.5 factors whenever validation accuracy did not improve for two consecutive epochs. The balanced class weight applied equally penalized the majority and minority classes. To avoid overfitting, an early stop (10 patients) was used, and the categorical cross-entropy loss function was used to measure network errors.

2.4 EXPERIMENTS SETTINGS

2.4.1 SETUP

The experiments were carried out on a workstation running Windows 10 with Intel Xeon Gold 6140 2 Processor 2.30 GHz 2.29 GHz, 256 GB memory, NVIDIA Quadro RTX 6000, dedicated GPU memory 24.0 GB, CUDA v10.0.130 Toolkit, and CuDNN v7.6.0.

2.4.2 DATASET

In this study publicly available CXR datasets and X-ray samples collected from Sardar Vallabhbhai Patel COVID Hospital, New Delhi, India, were used to train, test, and validate the proposed model. The details of the datasets used by the segmentation and classification network are as follows.

Dataset for Segmentation Network: The dataset utilized for the segmentation network was formed by combining three CXR datasets: (1) Montgomery County X-ray set [17]; (2) Shenzhen Hospital X-ray set [17]; and (3) JSRT dataset [26]. There are currently 138 CXRs and corresponding segmented lung masks in Montgomery County X-ray set, of which 58 are tuberculosis samples and 80 are normal samples. The Shenzhen Hospital X-ray set consists of 662 CXR samples, of which 326 are normal X-rays and 336 are tuberculosis X-ray. The corresponding manually segmented lung masks of Shenzhen Hospital X-ray set were collected from Stirenko et al. [29]. The JSRT dataset comprises 247 CX-ray samples including lung nodule and normal cases. The manual segmented masks of the JSRT dataset were collected from the Segmentation in Chest Radiographs (SCR) database [32]. Some of the collected CXR samples and their corresponding manually segmented lung masks are shown in Figure 2.2. After those three datasets had been combined, the dataset was

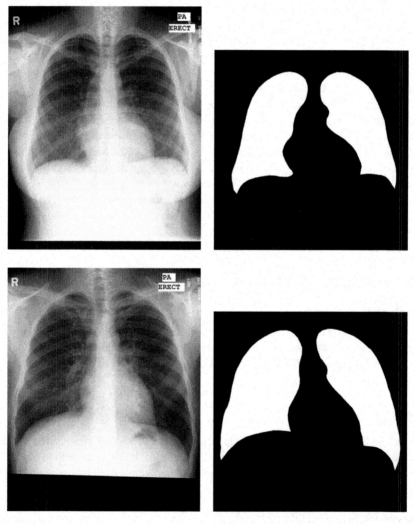

FIGURE 2.2   Some of the collected chest X-ray samples and their corresponding manually segmented lung masks.

TABLE 2.1
Dataset summary of the segmentation network

Dataset	Class	Chest X-ray samples	Segmented masks	Total
	Training			
Montgomery County X-ray set	Tuberculosis / normal	111	111	762
Shenzhen Hospital X-ray set / Stirenko et al.	Tuberculosis / normal	453	453	
JSRT dataset / SCR database	Nodule / normal	198	198	
	Validation			
Montgomery County X-ray set	Tuberculosis / normal	27	27	189
Shenzhen Hospital X-ray set / Stirenko et al.	Tuberculosis / normal	113	113	
JSRT dataset / Segmentation in Chest Radiographs (SCR) database	Nodule / normal	49	49	

randomly split into 80:20 ratios to train and validate our segmentation network, as shown in Table 2.1.

Dataset for Classification Network: COVID-19 Image Data Collection [10] and Figure 1 COVID-19 Chest X-ray Dataset Initiative [9] are currently the most popular open-source CXR datasets available for COVID-19 study and most of the researchers in this domain use COVID-19 X-ray samples from these sources. The COVID-19 Image Data Collection comprises 33 pneumonia and 218 COVID-19 samples; Figure 1 COVID-19 Chest X-ray Dataset Initiative comprises 27 COVID-19 samples. Normal and pneumonia CXR samples were collected from the Radiological Society of North America Pneumonia Detection Challenge dataset [20]. The dataset consists of 5,518 pneumonia and 8,066 normal X-ray samples. Further, 585 CXR samples of Indian patients were collected to test the proposed model.

Finally, the 14,692 samples collected were combined to train, validate, and test our classification network. Examples of the collected CXR samples of pneumonia, COVID-19, and normal patients are shown in Figure 2.3. Since coronavirus is a new disease, a limited number of coronavirus CXR samples are publicly available compared to normal or pneumonia samples and this leads to an imbalanced classification problem. To overcome this problem, a minority class over-sampling approach was adopted to generate more COVID-19 X-ray samples using 214 images from two [10,9] datasets, as proposed in our previous work [5] and 7,490 COVID-19 samples were obtained. The details of the dataset used for training and testing the classification network are summarized in Table 2.2.

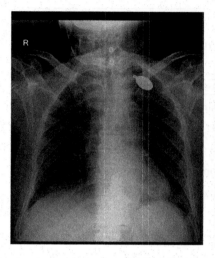
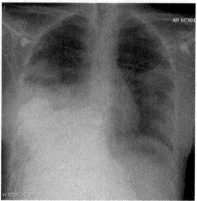
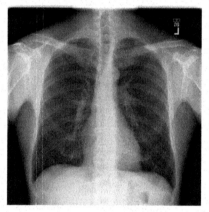

FIGURE 2.3   Chest X-ray samples of pneumonia, COVID-19, and normal patients.

TABLE 2.2
Dataset summary of the classification network

	Pneumonia	COVID-19	Normal	Total
Training	5,451	7,490	7,966	20,907
Testing	100	585	100	785

2.4.3   RESULTS AND ANALYSIS

The results and analysis of both networks – (1) segmentation and (2) classification – are presented in this section. Table 2.3 shows the performance of the segmentation network with different learning rates. We noticed that, during training, the learning

TABLE 2.3
Performance of the segmentation network with different learning rates

Learning rate	Train		Validation	
	Loss	Accuracy	Loss	Accuracy
lr = 1e⁻⁴	**0.3169**	**0.8814**	**0.1231**	**0.9654**
lr = 1e⁻⁴ and decay = 1e⁻⁵	0.3634	0.8619	0.2507	0.9475
lr = 1e⁻³ and decay = 1e⁻⁴	0.5779	0.7286	0.5558	0.7596
lr = 1e⁻³	4.2578	0.7284	3.8887	0.7587

rate has a direct influence on the model's efficiency. With learning $1e^{-4}$ the segmentation model achieved 96% validation accuracy and performed better than others. Figure 2.4 displays the original images along with the predicted lung masks and the segmented lungs obtained for different categorical classes using the segmentation network. From the extracted lung segments, it is visible that all the lung contours have distinctive and distinguishable features. In normal CXR images the lungs contour are well segmented but in the case of pneumonia and COVID-19, sometimes the segmented lung contour was deformed due to the excessive spread of infection with the virus across the lung.

Table 2.4 shows the performance of the classification network. The network achieved: (1) overall 98% test accuracy along with 99% recall and 99% precision for COVID-19 class; (2) 91% recall and 92% precision for pneumonia class; and (3) 95% recall and precision for normal class.

The categorical cross-entropy accuracy and loss plots of the classification network are shown in Figure 2.5. The validation curve fluctuates in accuracy and loss plots due to the limited number of COVID samples that are available in the dataset. In Figure 2.6, the confusion matrix illustrates the efficiency of the network. The model predicts only four samples as false-positive for the COVID-19 class. It misclassified five as pneumonia for the normal class, and for the pneumonia class, it misclassified four as COVID-19 and five as normal. It misclassified only three COVID-19 samples as pneumonia, because non-COVID pneumonia biomarkers can replicate the early biomarkers of the COVID-19 infection.

2.4.4 PERFORMANCE ASSESSMENT AGAINST INDIAN CORONAVIRUS PATIENT

To measure the two-stage DCNN network's performance in Indian coronavirus patients, 585 CXR samples of confirmed coronavirus patients were collected. We evaluated our proposed model's prediction results for those samples with the assistance of medical professionals from the Pune area, India. We observed that COVID-19 diagnosis sensitivity and the positive predicted value of our DCNN were about 99%. Some of the

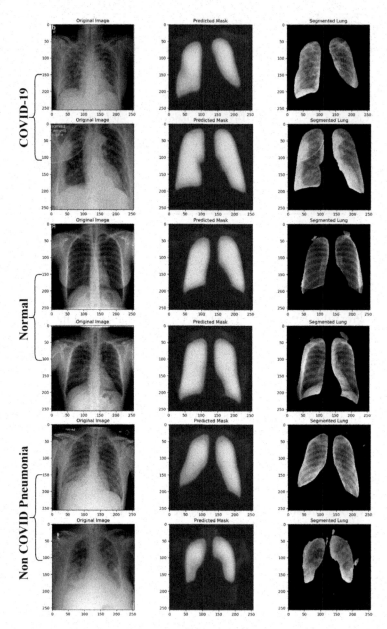

FIGURE 2.4 The original images along with the predicted lung masks and the segmented lungs obtained for different categorical classes using the segmentation network.

prediction results of the model are shown in Figure 2.7. The performance assessment results indicate the robustness of the model. It was also observed that the developed model helps to reduce diagnosis time and workload of the medical experts. As a result, medical professionals can focus on more critical patients. The medical experts' remarks on the output of the proposed two-stage DCNN network were as follows:

TABLE 2.4

Performance of the classification network on the test chest X-ray samples

	Precision	Recall	F1-score
COVID-19	99%	99%	99%
Normal	95%	95%	95%
Pneumonia	92%	91%	91%
Accuracy			98%
Macro average	95%	95%	95%
Weighted average	98%	98%	98%

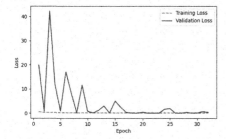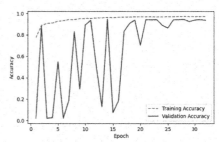

FIGURE 2.5   Categorical cross-entropy loss and accuracy plots of the classification network.

1. The proposed DCNN performed well in diagnosing coronavirus patients for the multiclass classification task.
2. The precision and sensitivity value of the DCNN were good in the context of Indian COVID-19 patients. This is in line with this model's purpose in which, in the present pandemic, we would not want to miss any COVID case.
3. The proposed low-cost tool can learn and differentiate the imaging biomarkers present in the CXR samples of normal, pneumonia, and coronavirus-infected patients.
4. The misclassification rate of the model is low. But still, further improvement is required, like training the model with more samples of all common thoracic diseases to make them more diversified.
5. The idea of examining only the infected area, i.e., lung contours in thoracic X-ray samples, and evaluating the model's feasibility for COVID-19 diagnosis was good.

2.4.5   PERFORMANCE COMPARISON WITH OTHER STATE-OF-THE-ART RESEARCH

Table 2.5 represents the performance comparison of the two-stage DCNN model with other state-of-the-art research. The proposed network achieved 98% overall test accuracy along with 99% sensitivity for coronavirus disease detection. To train the segmentation and classification network, 762 and 20,907 images (including synthetic images), respectively, were used. The same COVID-19 samples have been used in the studies listed in Tables 2.5 and 2.6. Category-wise precision and

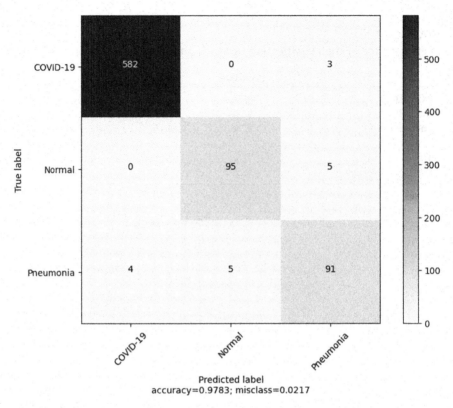

FIGURE 2.6   Confusion matrix of the classification network on the test dataset.

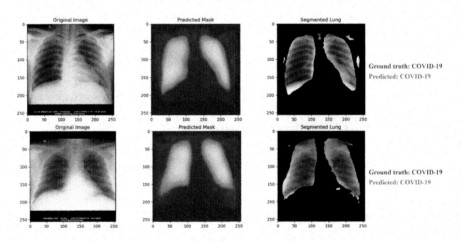

FIGURE 2.7   Some of the prediction results of the proposed model.

TABLE 2.5
Performance comparison with other state-of-the-art research

Methods	Model	Classes	Accuracy
Wang, Lin et al. [34]	COVID-Net	3	92.6%
Oh et al. [21]	FC-DenseNet103 + ResNet-18	3	91.9%
Apostolopoulos and Mpesiana [2]	VGG-19	3	93.48%
Hemdan et al. [14]	COVIDX-Net (VGG-19)	2	90%
Mangal et al. [19]	CovidAID	3	90.5%
Ozturk et al. [22]	DarkCovidNet	3	87.02%
Hira et al. [15]	Se-ResNeXt-50	3	96.99%
Our previous study [5]	Corona-Nidaan	3	95%
Proposed	U-Net + Corona-Nidaan	3	98%

TABLE 2.6
Category-wise precision and sensitivity values: comparison of proposed work with state-of-the-art studies

Study	Precision			Sensitivity		
	COVID-19	Normal	Pneumonia	COVID-19	Normal	Pneumonia
Wang, Lin et al. [34]	88.9%	91.3%	93.8%	80%	95%	91%
Oh et al. [21]	76.9%	95.7%	90.3%	100%	90%	93%
Mangal et al. [19]	96%	98%	86%	100%	74%	99%
Hemdan et al. [14]	83%	100%	–	100%	80%	–
Proposed	**99 %**	**95 %**	**92 %**	**99 %**	**95 %**	**91 %**

sensitivity value comparisons of the proposed work with other state-of-the-art studies are shown in Table 2.6. Wang, Lin et al. [34] designed a new DCNN architecture, namely COVID-Net, for coronavirus disease diagnosis. Their proposed network achieves 92.6% accuracy but suffers from false-negative results. The diagnostic sensitivity of coronavirus disease is lower (80%) than this model (99%), and the precision value of COVID-Net is also relatively lower than this model for normal (91.3%) and COVID-19 (88.9%) (Table 2.6). Oh et al. [21] proposed a two-stage DCNN model and reported 91.9% test accuracy. COVID-19 diagnostic sensitivity of the model is close to this model, but for normal patients, the sensitivity is lower than this model. The precision value for pneumonia (90.3%) and COVID-19 (76.9%) classes is lower than this model (Table 2.6). They collected a minimal number of CXR samples to train the model. Apostolopoulos and Mpesiana [2], and Hemdan et al. [14] utilized transfer learning for diagnosis of coronavirus, and both reported that VGG-19 performed better in their experiments with pretrained CNNs. However, they trained their network with minimal CXR samples and achieved 93.48% and 90% accuracy, respectively. Mangal et al. [19] obtained 90.5% overall test accuracy by utilizing transfer learning in their proposed CovidAID deep learning model. The sensitivity of their model for COVID-19 detection was closest to this study, but they suffered from false-negative results for normal CXR. Ozturk et al. [22] designed DarkCovidNet, inspired by DarkNet DCNN, and trained their model with a limited number of non-COVID images and achieved 87.02% accuracy. They utilized the under-sampling technique to tackle the imbalanced classification problem. The F1-score, sensitivity, and precision of their model were lower than this model for three-class classification. Hira et al. [15] used 224 coronavirus, 504 normal, and 700 bacterial pneumonia CXR samples to train nine pre-trained CNNs and attained 96.99% accuracy. The model trained with minimal non-COVID samples, and the average sensitivity (94.67%), precision (87.36%), and F1-score (90.86%) were also lower than this study. The proposed model's overall accuracy (98%) was better than our previous work [5] (95%); we were also able to achieve 99% sensitivity for COVID-19 diagnosis and 95%, 99% precision value for the normal, COVID-19 class, respectively. This work's main objective was not to miss a single COVID-19 case so that this model can be promoted for rapid testing, but not as an alternative to RT-PCR.

2.5 CONCLUSION

This chapter presents a two-stage DCNN model to diagnose coronavirus disease by utilizing CXR samples. The proposed model is divided into two parts: (1) segmentation network; and (2) classification network. From the early success of the U-Net model and the Corona-Nidaan DCNN, we considered them as a backbone of the segmentation and classification network. The findings in this study can be useful in making decisions in clinical practice for doctors, radiologists, and researchers. The following conclusions can be made based on the findings of the experiments:

1. The proposed two-stage DCNN model achieved overall 98% test accuracy along with 99% sensitivity for COVID-19 detection.
2. The sensitivity of COVID-19 detection depends on how efficiently we extract lung segments from CXR samples.
3. Using deep learning and automated methods reawakens the relevance of CXR samples, and they can be used for learning the hallmarks of coronavirus disease.
4. The false-negative and positive rate of the proposed model is comparatively lower than in early studies.
5. The proposed model has been stably trained with limited datasets and achieves state-of-the-art performance.
6. The sensitivity of the model is better than RT-PCR for COVID-19 diagnosis, so it can be used as a reliable tool to enable rapid testing of coronavirus patients, but not as an alternative to RT-PCR.

We will enhance the proposed DCNN in our future work to diagnose all kinds of lung disease and analyze the severity of those diseases by collecting more CXR samples.

ACKNOWLEDGMENTS

This research is supported by the Defence Institute of Advanced Technology, DRDO Lab, Ministry of Defence, India, and the Indian National Academy of Engineering. The authors would like to thank Sardar Vallabhbhai Patel COVID Hospital, New Delhi, India, for facilitating testing of the proposed model by providing the Indian patient X-ray samples. The authors would like to thank NVIDIA for the GPU grant for carrying out deep learning-based research work.

REFERENCES

[1] Ahuja, Aastha, and Sonia Bhaskar. "Experts explain the different Covid-19 tests: Rapid antigen vs RT-PCR test, which is better?" https://swachhindia.ndtv.com/experts-expl ain-the-different-covid-19-tests-rapid-antigen-vs-rt-pcr-test-which-is-better-48040/ (2020).

[2] Apostolopoulos, Ioannis D., and Tzani A. Mpesiana. "Covid-19: Automatic detection from X-ray images utilizing transfer learning with convolutional neural networks." *Physical and Engineering Sciences in Medicine* 43, no. 2 (2020): 635–640.

[3] Bai, Harrison X., Ben Hsieh, Zeng Xiong, Kasey Halsey, Ji Whae Choi, Thi My Linh Tran, Ian Pan et al. "Performance of radiologists in differentiating COVID-19 from non-COVID-19 viral pneumonia at chest CT." *Radiology* 296, no. 2 (2020): E46–E54.

[4] Bejnordi, Babak Ehteshami, Mitko Veta, Paul Johannes Van Diest, Bram Van Ginneken, Nico Karssemeijer, Geert Litjens, Jeroen AWM Van Der Laak et al. "Diagnostic assessment of deep learning algorithms for detection of lymph node metastases in women with breast cancer." *Jama* 318, no. 22 (2017): 2199–2210.

[5] Chakraborty, Mainak, Sunita Vikrant Dhavale, and Jitendra Ingole. "Corona-Nidaan: lightweight deep convolutional neural network for chest X-ray based COVID-19 infection detection." *Applied Intelligence* 51, no. 5 (2021): 3026–3043.

[6] Chakraborty, Mainak, Alik Pramanick, and Sunita Vikrant Dhavale. "MobiSamadhaan—Intelligent vision-based smart city solution." In *International Conference on Innovative Computing and Communications*, pp. 329–345. Springer, Singapore, 2021.

[7] Chakraborty, Mainak, Alik Pramanick, and Sunita Vikrant Dhavale. "Two-stream mid-level fusion network for human activity detection." In *International Conference on Innovative Computing and Communications*, pp. 331–343. Springer, Singapore, 2021.

[8] Chan, Jasper Fuk-Woo, Cyril Chik-Yan Yip, Kelvin Kai-Wang To, Tommy Hing-Cheung Tang, Sally Cheuk-Ying Wong, Kit-Hang Leung, Agnes Yim-Fong Fung et al. "Improved molecular diagnosis of COVID-19 by the novel, highly sensitive and specific COVID-19-RdRp/Hel real-time reverse transcription-PCR assay validated in vitro and with clinical specimens." *Journal of Clinical Microbiology* 58, no. 5 (2020).

[9] Wang, Linda, Alexander Wong, Zhong Qiu Lin, James Lee, Paul McInnis, Audrey Chung, Matt Ross, Blake van Berlo, and Ashkan Ebadi. "Figure 1 COVID-19 chest X-ray dataset initiative." https://github.com/agchung/Figure1-COVID-chestxray-dataset

[10] Cohen, Joseph Paul, Paul Morrison, Lan Dao, Karsten Roth, Tim Q. Duong, and Marzyeh Ghassemi. "COVID-19 image data collection: Prospective predictions are the future." arXiv preprint arXiv:2006.11988 (2020).

[11] Gaál, Gusztáv, Balázs Maga, and András Lukács. "Attention u-net based adversarial architectures for chest X-ray lung segmentation." arXiv preprint arXiv:2003.10304 (2020).

[12] Gordienko, Yu, Peng Gang, Jiang Hui, Wei Zeng, Yu Kochura, Oleg Alienin, Oleksandr Rokovyi, and Sergii Stirenko. "Deep learning with lung segmentation and bone shadow exclusion techniques for chest X-ray analysis of lung cancer." In *International Conference on Computer Science, Engineering and Education Applications*, pp. 638–647. Springer, Cham, 2018.

[13] Gulshan, Varun, Lily Peng, Marc Coram, Martin C. Stumpe, Derek Wu, Arunachalam Narayanaswamy, Subhashini Venugopalan et al. "Development and validation of a deep learning algorithm for detection of diabetic retinopathy in retinal fundus photographs." *Jama* 316, no. 22 (2016): 2402–2410.

[14] Hemdan, Ezz El-Din, Marwa A. Shouman, and Mohamed Esmail Karar. "Covidx-net: A framework of deep learning classifiers to diagnose COVID-19 in X-ray images." arXiv preprint arXiv:2003.11055 (2020).

[15] Hira, Swati, Anita Bai, and Sanchit Hira. "An automatic approach based on CNN architecture to detect COVID-19 disease from chest X-ray images." *Applied Intelligence* (2020): 1–26.

[16] Huang, Chaolin, Yeming Wang, Xingwang Li, Lili Ren, Jianping Zhao, Yi Hu, Li Zhang et al. "Clinical features of patients infected with 2019 novel coronavirus in Wuhan, China." *The Lancet* 395, no. 10223 (2020): 497–506.

[17] Jaeger, Stefan, Sema Candemir, Sameer Antani, Yì-Xiáng J. Wáng, Pu-Xuan Lu, and George Thoma. "Two public chest X-ray datasets for computer-aided screening of pulmonary diseases." *Quantitative Imaging in Medicine and Surgery* 4, no. 6 (2014): 475.

[18] Lakhani, Paras, and Baskaran Sundaram. "Deep learning at chest radiography: Automated classification of pulmonary tuberculosis by using convolutional neural networks." *Radiology* 284, no. 2 (2017): 574–582.

[19] Mangal, Arpan, Surya Kalia, Harish Rajgopal, Krithika Rangarajan, Vinay Namboodiri, Subhashis Banerjee, and Chetan Arora. "CovidAID: COVID-19 detection using chest X-ray." arXiv preprint arXiv:2004.09803 (2020).

[20] Radiological Society of North America (RSNA). "Rsna pneumonia detection challenge dataset." www.kaggle.com/c/rsna-pneumonia-detection-challenge/data (2019).

[21] Oh, Yujin, Sangjoon Park, and Jong Chul Ye. "Deep learning COVID-19 features on CXR using limited training data sets." *IEEE Transactions on Medical Imaging* 39, no. 8 (2020): 2688–2700.

[22] Ozturk, Tulin, Muhammed Talo, Eylul Azra Yildirim, Ulas Baran Baloglu, Ozal Yildirim, and U. Rajendra Acharya. "Automated detection of COVID-19 cases using deep neural networks with X-ray images." *Computers in Biology and Medicine* 121 (2020): 103792.

[23] Rajpurkar, Pranav, Awni Y. Hannun, Masoumeh Haghpanahi, Codie Bourn, and Andrew Y. Ng. "Cardiologist-level arrhythmia detection with convolutional neural networks." arXiv preprint arXiv:1707.01836 (2017).

[24] Rajpurkar, Pranav, Jeremy Irvin, Kaylie Zhu, Brandon Yang, Hershel Mehta, Tony Duan, Daisy Ding et al. "Chexnet: Radiologist-level pneumonia detection on chest x-rays with deep learning." arXiv preprint arXiv:1711.05225 (2017).

[25] Ronneberger, Olaf, Philipp Fischer, and Thomas Brox. "U-net: Convolutional networks for biomedical image segmentation." In *International Conference on Medical Image Computing and Computer-Assisted Intervention*, pp. 234–241. Springer, Cham, 2015.

[26] Shiraishi, Junji, Shigehiko Katsuragawa, Junpei Ikezoe, Tsuneo Matsumoto, Takeshi Kobayashi, Ken-ichi Komatsu, Mitate Matsui, Hiroshi Fujita, Yoshie Kodera, and Kunio Doi. "Development of a digital image database for chest radiographs with and without a lung nodule: Receiver operating characteristic analysis of radiologists' detection of pulmonary nodules." *American Journal of Roentgenology* 174, no. 1 (2000): 71–74.

[27] Simonyan, Karen, and Andrew Zisserman. "Two-stream convolutional networks for action recognition in videos." arXiv preprint arXiv:1406.2199 (2014).

[28] Sohrabi, Catrin, Zaid Alsafi, Niamh O'Neill, Mehdi Khan, Ahmed Kerwan, Ahmed Al-Jabir, Christos Iosifidis, and Riaz Agha. "World Health Organization declares global emergency: A review of the 2019 novel coronavirus (COVID-19)." *International Journal of Surgery* 76 (2020): 71–76.

[29] Stirenko, Sergii, Yuriy Kochura, Oleg Alienin, Oleksandr Rokovyi, Yuri Gordienko, Peng Gang, and Wei Zeng. "Chest X-ray analysis of tuberculosis by deep learning with segmentation and augmentation." In *2018 IEEE 38th International Conference on Electronics and Nanotechnology (ELNANO)*, pp. 422–428. IEEE, 2018.

[30] Tandon, Prakash N. "COVID-19: Impact on health of people and wealth of nations." *The Indian Journal of Medical Research* 151, no. 2–3 (2020): 121.

[31] Udugama, Buddhisha, Pranav Kadhiresan, Hannah N. Kozlowski, Ayden Malekjahani, Matthew Osborne, Vanessa YC Li, Hongmin Chen, Samira Mubareka, Jonathan B. Gubbay, and Warren C.W. Chan. "Diagnosing COVID-19: The disease and tools for detection." *ACS Nano* 14, no. 4 (2020): 3822–3835.

[32] Van Ginneken, Bram, Mikkel B. Stegmann, and Marco Loog. "Segmentation of ana-tomical structures in chest radiographs using supervised methods: A comparative study on a public database." *Medical Image Analysis* 10, no. 1 (2006): 19–40.

[33] Wang, Li, and Dennis Sng. "Deep learning algorithms with applications to video analytics for a smart city: A survey." arXiv preprint arXiv:1512.03131 (2015).

[34] Wang, Linda, Zhong Qiu Lin, and Alexander Wong. "Covid-net: A tailored deep con-volutional neural network design for detection of COVID-19 cases from chest X-ray images." *Scientific Reports* 10, no. 1 (2020): 1–12.

[35] Wang, Xiaosong, Yifan Peng, Le Lu, Zhiyong Lu, Mohammadhadi Bagheri, and Ronald M. Summers. "Chestx-ray8: Hospital-scale chest X-ray database and benchmarks on weakly-supervised classification and localization of common thorax diseases." In *Proceedings of the IEEE Conference on Computer Vision and Pattern Recognition*, pp. 2097–2106. IEEE, 2017.

[36] WHO. "Coronavirus disease (COVID-19)." www.who.int/emergencies/diseases/novel-coronavirus-2019/question-and-answers-hub/q-a-detail/coronavirus-disease-covid (2020).

[37] WHO. "Coronavirus disease (COVID-19): How is it transmitted?" www.who.int/emergencies/diseases/novel-coronavirus-2019/question-and-answers-hub/q-a-detail/coronavirus-disease-covid-19-how-is-it-transmitted (2020).

[38] WHO. "Who coronavirus disease (COVID-19) dashboard." https://covid19.who.int/, (2020).

[39] Wu, Fan, Su Zhao, Bin Yu, Yan-Mei Chen, Wen Wang, Zhi-Gang Song, Yi Hu et al. "A new coronavirus associated with human respiratory disease in China." *Nature* 579, no. 7798 (2020): 265–269.

3 White Blood Cell Classification Using Conventional and Deep Learning Techniques

A Comparative Study

P. Pandiyan
KPR Institute of Engineering and Technology, Coimbatore,
Tamil Nadu, India

Rajasekaran Thangaraj
KPR Institute of Engineering and Technology, Coimbatore,
Tamil Nadu, India

S. Srinivasulu Raju
VR Siddhartha Engineering College, Vijayawada,
Andhra Pradesh, India

Vishnu Kumar Kaliappan
KPR Institute of Engineering and Technology, Coimbatore,
Tamil Nadu, India

B. Lalitha
KPR Institute of Engineering and Technology, Coimbatore,
Tamil Nadu, India

3.1 INTRODUCTION

Blood plays a significant role in life. The functionalities of many organs in the body depend on healthy blood. The assessment of cells available in the blood determines the healthiness of the blood. Generally, blood is a combination of solid and liquid parts. The liquid portion is referred to as plasma, and is made up of protein, water, and salts. The solid part of the blood consists of white blood cells (WBCs), platelets, and red blood cells (RBCs) [1]. There are no nuclei in RBCs, whereas WBCs have

DOI: 10.1201/9781003230540-3

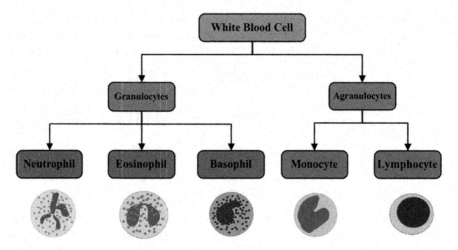

FIGURE 3.1 Classification of white blood cells.

cytoplasm and a nucleus. The RBC constitutes about 42–45% of blood volume, whereas the WBC constitutes about 1% of blood volume.

Each blood cell has different functionalities for organs of the body. WBCs are mainly accountable for the immunity development of the body, which acts as an opponent to disease-causing foreign elements. WBCs are generated from hematopoietic stem cells present in the bone marrow. Depending upon nucleus structures, WBCs are classified into five categories: basophil (0–1%), lymphocyte (20–45%), eosinophil (1–5%), neutrophil (50–70%), and monocyte (2–10%) [2]. These are again clustered into two major categories, namely granulocytes and agranulocytes, depending upon the presence of granules in the cytoplasm, shown in Figure 3.1. Granulocytes are WBCs with visible granules, whereas agranulocytes do not have visible granules when examined under a microscope [3]. The content of these WBC types reflects the health of the patient.

Neutrophils have two subpopulations that are unequal in volume and defend against fungal and bacterial infections. They are called neutrophil cagers and neutrophil killers. Eosinophils are responsible for parasitic infections, allergies, disease of the spleen, collagen disease, and the central nervous system. Basophils rise in response to antigens and allergic responses by discharging histamine to enlarge the blood vessels. Monocytes are helpful in destroying attackers but also enable repair and healing. Lymphocytes aid immune blood cells to associate with other external organisms, viz. antigens and microorganisms, to eliminate them from the body.

Medical professionals generally use this percentage of WBCs to identify the significance of blood disease. The contribution of WBCs in the blood is essential in creating immunity in the human body. Therefore, the classification of WBCs plays a vital role in the medical field. In order to assess their quantity in standard proportions or quantities, it requires exact identification and counting. Identification of different WBC types can be of further help in discovering abnormal conditions. The quantitative and qualitative assessment of WBCs discloses more information about the patient's health condition. For instance, it is used to examine the patient's health, such as immune system disorders, leukaemia and cancerous cells, allergies, infection, tissue injury, and bronchogenic conditions [4]. Similarly, an abnormal RBC count may indicate diseases such as lung diseases and kidney tumours.

In conventional methods, the identification of WBC needs a laboratory setup where collected blood samples are stained using reagents and observed under a microscope by an expert manually. This type of process leads to minimal observational errors, less reliability, and a time-consuming and tedious process. In order to reduce these issues, it is imperative to ensure highly experienced haematologists are available in the laboratory and produce the results at a high level of consistency. In some cases, human specialists may become tired due to several hours of observation and the inaccurate identification of different WBCs. Therefore, a computer-based system is required to identify and find the blood cell count. Modern analytical equipment is capable of examining 1000 samples/day for automatic blood cell analysis.

The manual and conventional methods of differentiating WBCs use many image pre-processing approaches before performing the feature extraction process. This process takes more time and is more complex [5]. The hand-crafted engineering technique uses statistical features to represent blood cells and feeds them into the classifiers to predict the output. Nevertheless, manual feature engineering is challenging because it depends on the accuracy of manually extracted information, which leads to high processing time and low classification performance. Henceforth, accurate prediction of WBCs is essential to classify the blood cells automatically, anticipating that the risk related to blood syndromes will be lowered.

Automatic blood cell identification, classification, counting, and analysis through a computer vision approach can enable the blood test to be performed accurately and efficiently. The automated blood cell analysis system comprises three stages, namely blood cell image segmentation (BCIS), feature extraction (FE), and blood cell classification (BCC).

Deep learning models have recently become popular and comprise several deep network layers (deep convolutional neural network: DCNN). Deep learning based on CNN is the technique employed most nowadays for image classification to overcome the shortcomings faced by traditional approaches. The DCNN will solve the problems of low accuracy, weight matrices, and high processing time [6]. This approach efficiently manages non-linear relationships between image databases and classes.

This review paper discusses the existing conventional and deep learning approaches towards classifying WBCs and their performance. Section 3.4 explains the challenges faced in executing the classification of WBCs. Finally, a conclusion is drawn in the last section.

3.2 CONVENTIONAL METHODS USED IN BLOOD CELL CLASSIFICATION

There are several automated cell morphology systems (ACMS), such as Hema-CAM, MEDICA EasyCell Assistant, and Cella-Vision [2]. These systems use the conventional identification approach, as illustrated in Figure 3.2. In conventional WBCs, the prediction system consists of pre-processing, segmentation, feature extraction, feature selection, and classification.

In the pre-processing stage, de-noising, as well as colour correction processes, is carried out. Commonly used image enhancement algorithms such as histogram

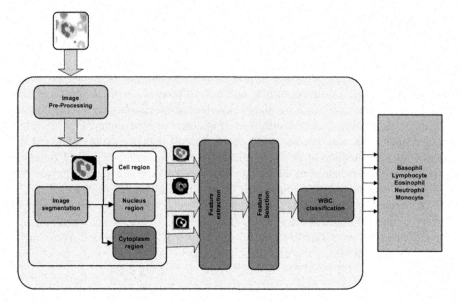

FIGURE 3.2   Conventional white blood cell (WBC) classification system [2].

matching and average filtering are employed for enhancing blood smear images. The segmentation phase plays a significant role in the identification/classification system. The overall system accuracy relies on this phase [7]. The segmentation algorithms used in blood cell identification include support vector machine (SVM), an online trained neural network [8]. There are different features, viz. colour, morphology, texture, etc., for both cytoplasm and nuclei regions in the feature extraction stage.

K-nearest neighbour, naive Bayes, SVM, multilayer perceptron, and ensemble methods in machine learning are the conventional methods used for detection of WBC.

In the past decade, more research has been performed to identify WBCs. A summary of the earlier pathological identification of the WBC system is given in Table 3.1. This table gives the details such as dataset size, features used, types of classifier, and prediction accuracy.

Summary of the previous work carried out for classification of WBC attains higher accuracy only when environmental conditions are strictly controlled as mentioned in Su [16] or small datasets are utilized for classification, as in [9,13,14]. Rezatofighi and Soltanian-Zadeh [10] proposed colour adjustment in the input image as a pre-processing step to enhance classification accuracy. Even so, some challenges have to be considered to improve accuracy. The literature review reveals that prediction accuracy can be enhanced by increasing the efficiency of the segmentation algorithms. But any under- or over-segmentation errors have a negative effect on overall accuracy. The feature extraction stage also plays a significant role in automatic classification. Putzu et al. [12] reported that the morphological features are prone to creating errors in the segmentation stage. The WBC features were extracted from red, green, and blue (RGB) colour components individually by Prinyakupt and Pluempitiwiriyawej [8]. Furthermore, the ratios of statistical features of RGB colour components are also considered, increasing the complexity of the feature extraction phase. A multi-step classification approach was followed in [16, 30, 32].

TABLE 3.1
Comparison table showing the performance of conventional approach for classifying white blood cells

Reference	Year published	Dataset size	Classifier used	Classification accuracy
Wang et al. [9]	2016	60	SVM	90%
Rezatofighi and Soltanian-Zadeh [10]	2011	254	ANN + SVM	96%
Prinyakupt and Pluempitiwiriyawej [8]	2015	1078	Linear and Naïve Bayes	96%
Ramesh et al. [11]	2012	2172	LDA	93.9%
Putzu et al. [12]	2014	33	SVM	93%
Nazlibilek et al. [13]	2014	12	MLP	95%
Mathur et al. [14]	2013	267	Naïve Bayes	92.7%
Ghosh et al. [15]	2016	487	Fuzzy classifier	96%
Su et al. [16]	2014	450	SVMs, HCNN, MLPs	99.1%
Adjouadi et al. [17]	2005	100	SVMs	87%
Ghosh et al. [18]]	2010	150	Naïve Bayes	83.2%
Habibzadeh et al. [19]	2013	140	SVMs and K-PCA	76–84%
Schneider et al. [20]	2015	7500	Optical neural network	89%
Ravikumar [21]	2016	–	ELMs and fast RVM	82.45–92.6%
Zhao et al. [22]	2017	288	Random forest	–
Al-Dulaimi et al. [23]	2018	460	SVMs and classification tree	96.13%
Al-Dulaimi et al. [24]	2018	460	SVMs and classification tree	97.23%
Ko et al. [25]	2011	300	RF	72.5%
Ravikumar and Shanmugam [26]	2015	–	RVM	81–91%
Nassar et al. [27]	2019	85	AdaBoost, GB, KNN, RF, and SVM	78–97% (F1 Score)
JS [28]	2020	1030	SVM with different kernels	57–62.62%
Vogado et al. [29]	2016	735	K-means clustering algorithm	99.15
Kumar and Vasuki [30]	2017	70	Multi-SVM	90%
Nasir et al. [31]	2011	100	K-means clustering algorithm	99.02%
Sajjad et al. [32]	2016	1030	Multi-class ensemble-SVM	98.8%

SVM, support vector machine; ANN, artificial neural network; LDA, linear discriminant analysis; MLP, multilayer perceptron; HCNN, heterogeneous convolutional neural network; K-PCA, K-principal component analysis; ELM, extreme learning machine; RVM, relevance vector machine; GB, gradient boosting; K-NN, K-nearest neighbors; RF, random forest.

3.3 DEEP LEARNING MODELS USED IN BLOOD CELL CLASSIFICATION

3.3.1 DATASETS

Datasets are essential for researching deep learning algorithms. The datasets used for classifying the blood cells were BCCD, Cellavission, ALL-IDB, PASCAL VOC (2007 and 2012), and MS COCO datasets. Table 3.2 gives a summary of the WBC image datasets.

3.3.2 DEEP LEARNING MODELS

Deep learning using CNN is currently the best choice for medical imaging applications such as detection and classification. In this section, the architecture of deep learning models used for classifying blood cells, such as fully trained CNN, transfer learning, and CNN as a feature generator, is discussed. The deep learning approach works on raw images that extract the features through their inherent characteristics. The most widely used deep learning methods in medical image analysis are stacked autoencoders, AlexNet, Resnet50, DenseNet201, and GoogleNet. A comparison chart describing the deep learning model adopted to classify WBCs is shown in Table 3.3 in terms of the year published, CNN model employed, and accuracy achieved.

TABLE 3.2
Summary of white blood cell image datasets

Reference	Dataset/database used
Adjouadi et al. [17]	BCCD
Ghosh et al. [15]	MMCH
Rezatofighi and Soltanian-Zadeh [10]	HBRCH
Su et al. [16]	Cella vision database
Schneider et al. [20]	Flow cytometer databases
Ravikumar [21]	Hospital database
Zhao et al. [22]	Cellavision, ALL-IDB and Jiashan
Habibzadeh et al. [19]	Hospital database
Al-Dulaimi et al. [23, 24]	Cellavision-Database (2011]) All-IDB (2005) and Wadsworth Center
Kutlu et al. [33]	BCCD and LISC
Banik et al. [34]	BCCD, ALL-IDB2, JTSC and CellaVision
Nazlibilek et al. [35]	www.kanbilim.com
Vogado et al. [29]	LL-IDB2, SegBlood and Leukocytes
Vincent et al. [36]	AML and ALL
Imran Razzak and Naz [37]	ALL-IDB
Yu et al. [38]	Drum Tower Hospital, Nanjing University Medical School
Tran et al. [39]	ALL-IDB1
Fan et al. [40]	BCISC and LISC

TABLE 3.3

Comparison table showing performance of deep learning algorithm applied to classify white blood cells

Reference	Year	Model/method used		Accuracy (%)	
Nazlibilek et al. [13]	2014	Neural network		95	
Liang et al. [41]	2018	CNN and RNN		90.79	
Bani-Hani et al. [42]	2018	CNN and genetic algorithm		91.00	
Hegde et al. [43]	2019	CNN and LBP		99.00	
Sharma et al. [44]	2019	CNN and data augmentation approach		87.00	
Wang et al. [45]	2018	PatternNet-fused ensemble of CNN		99.90	
Kutlu et al. [33]	2020	R-CNN – AlexNet		92.00	
		Fast R-CNN – AlexNet		92.00	
		Faster R-CNN – AlexNet		100	
		SSD – AlexNet		95	
		YOLO v3		96	
Banik et al. [34]	2020	Fusion-based CNN model on BCCD dataset		99.42	
		Fusion-based CNN model on ALL-IDB–2 dataset		98.61	
		Fusion-based CNN model on JTSC dataset		97.57	
		Fusion-based CNN model on CellaVision dataset		98.86	
H. Mohamed et al. [46]	2020	VGG-16	VGG-19	73.64	71.93
		ResNet-50		52.42	
		DenseNet-121	DenseNet-169	83.53	86.04
		Inception-V3	Inception ResNet-V2	77.11	78.48
		Xception		71.16	
		1.0 MobileNet-224	Mobile NasNet	79.77	74.06
Vincent et al. [36]	2015	Two-step neural network		97.7 %	
Imran Razzak and Naz [37]	2017	Faster-region CNN		98.6	
Yu et al. [38]	2017	CNN model		88.5	
Choi et al. [47]	2017	Dual-stage CNN model		97.06	
Rehman et al. [48]	2018	CNN model		97.78	
Alom et al. [49]	2018	Inception recurrent residual CNN		99.4	
Tran et al. [39]	2018	Deep learning semantic segmentation		94.93	
Wang et al. [50]	2018	PatternNet-fused ensemble of CNNs		98.67	
Shahin et al. [2]	2019	DeCA AlexNet		79.8	
		DeCA overfeat		76.6	
		DeCA VGG		72.3	
		Fine-tuned AlexNet		91.2	
		Fine-tuned LeNet		84.9	
Wang et al. [51]	2019	SSD		90.09	
		YOLO V3		89.36	
Jha and Dutta [52]	2019	Hybrid segmentation + Chrono-SCA-DCNN		98.7	
Pandey et al. [53]	2020	Target independent generative domain adaptation		75.8	

(continued)

TABLE 3.3 Cont.
Comparison table showing performance of deep learning algorithm applied to classify white blood cells

Reference	Year	Model/method used	Accuracy (%)
Basnet et al. [54]	2020	DCCN: enhanced loss function with regularization and weighted loss	98.92%
Patil et al. [55]	2020	Canonical correlation analysis	95.89
Reena and Ameer [56]	2020	Semantic segmentation based on deep learning	98.87

CNN, convolutional neural network; RNN, recurrent neural network; LBP, local binary pattern; R-CCN, region-based CNN; SSD, single shot detector; DCNN, deep convolutional neural network.

3.3.3 PERFORMANCE EVALUATION METRICS

The performance of the deep learning model is evaluated using precision, accuracy, F1-score, and recall. The expressions for the performance metrics are given in Equations 3.1–3.4.

$$Accuracy = \frac{TP+TN}{TP+TN+FP+FN} \tag{1}$$

$$Precision = \frac{TP}{TP+FP} \tag{2}$$

$$Recall = \frac{TP}{TP+FP} \tag{3}$$

$$F1\ score = 2 \times \frac{Recall \times Precision}{Recall + Precision} \tag{4}$$

where TP = true positive, FP = false positive, TN = true negative, and FN = false negative. Table 3.4 shows a comparison chart illustrating the deep learning model employed to classify WBCs in terms of performance evaluation metrics.

3.4 CHALLENGES AND FUTURE SCOPE

The accuracy of the WBC classification model is highly dependent on two stages, namely, image segmentation and feature extraction. The classification model needs to consider the following problems: size changes, location, different relative size and position of the nucleus, rotation of blood cells, and different maturation stages. The challenges faced in implementing the WBC algorithm are as follows:

1. The different types of staining process lead to changes in nucleus colorations, background, and cytoplasm and may lead to cell distortion.

2. The nucleus present in WBCs may appear at different sizes (small or large)/ locations in different images.

3. WBCs attain the maturation stage from the primitive blast stage in the blood, which causes changes in nucleus shape, cytoplasm, cell size, and position. These stages of growth are complex because minor inter-class differences exist between the successive stages.

4. WBC images may be viewed at arbitrary angles, and their appearance may differ when monitored from different angles. Therefore, the classification model should incorporate this issue.

5. Most of the images of the datasets are taken in perfect lighting conditions, but conditions slightly differ in real time and lead to a different output.

6. The diversity of target datasets and selecting the best architecture suitable for the target class are significant.

7. There is no standardized computer vision technology available for WBC classification.

8. Methods used to detect WBCs failed to inform about the severity of the disease and how to rectify it.

9. CNN/DCNN models with smaller datasets produce greater prediction accuracy, but that is not a reliable and trustworthy result.

10. A higher computational cost is required for running the CNN/DCNN architecture on central processing units (CPUs) compared to GPUs.

11. The CNN architecture does not give the best data classification by incorporating multiple convolutions.

12. As of now, no reliable and flexible architecture can predict and classify the complex and challenging WBC classification.

The future scope of WBC classification is as follows. In order to achieve the excellent performance of deep learning models a fusion of low- and high-level pre-trained model layers needs to be performed. The convergence time of deep learning models can be reduced by decreasing the number of parameters involved and improving prediction accuracy.

TABLE 3.4

Comparison table showing performance of deep learning model applied to classify white blood cells in terms of performance evaluation metrics

Reference	Model	Precision	Sensitivity	Specificity	Recall
Imran Razzak and Naz [37]	Faster-region CNN	98.3	95.78	98.99	95.93
Choi et al. [47]	Dual-stage CNN model	97.13	96.8	97.1	97.06

CNN, convolutional neural network.

3.5 CONCLUSIONS

In this work, reviews related to classifying WBCs using machine learning, shallow and deep learning architectures were reported and analyzed. The usage of public and private datasets and a variety of pre-trained models were also analyzed. From the review, the highest recognition accuracy can be obtained using deep learning architecture compared to other well-known techniques such as image processing techniques, machine learning, and neural networks to classify the WBC through peripheral blood smear images. Furthermore, Faster RCNN – AlexNet gives the best accuracy and greater prediction accuracy.

REFERENCES

[1] Turgeon, M. L. *Clinical hematology: Theory and procedures*. Guilin, China: Lippincott Williams and Wilkins, 2005.

[2] Shahin, A. I., Y. Guo, K. M. Amin, and A. A. Sharawi. "White blood cells identification system based on convolutional deep neural learning networks." *Computer Methods and Programs in Biomedicine* 168 (2019): 69–80.

[3] Mathur, Atin, Ardhendu S. Tripathi, and Manohar Kuse. "Scalable system for classification of white blood cells from Leishman stained blood stain images." *Journal of Pathology Informatics* 4, Suppl (2013).

[4] Shafique, Sarmad, and Samabia Tehsin. "Computer-aided diagnosis of acute lymphoblastic leukaemia." *Computational and Mathematical Methods in Medicine* 2018 (2018).

[5] Janowczyk, Andrew, and Anant Madabhushi. "Deep learning for digital pathology image analysis: A comprehensive tutorial with selected use cases." *Journal of Pathology Informatics* 7 (2016).

[6] Spampinato, Concetto, Simone Palazzo, Daniela Giordano, Marco Aldinucci, and Rosalia Leonardi. "Deep learning for automated skeletal bone age assessment in X-ray images." *Medical Image Analysis* 36 (2017): 41–51.

[7] Wu, Jianhua, Pingping Zeng, Yuan Zhou, and Christian Olivier. "A novel color image segmentation method and its application to white blood cell image analysis." In *2006 8th international Conference on Signal Processing*, vol. 2. IEEE, 2006.

[8] Prinyakupt, Jaroonrut, and Charnchai Pluempitiwiriyawej. "Segmentation of white blood cells and comparison of cell morphology by linear and naïve Bayes classifiers." *Biomedical Engineering Online* 14, no. 1 (2015): 1–19.

[9] Wang, Q., L. Chang, M. Zhou, Q. Li, H. Liu, and F. Guo. "A spectral and morphologic method for white blood cell classification." *Optics and Laser Technology* 84 (2016): 144–148.

[10] Rezatofighi, Seyed Hamid, and Hamid Soltanian-Zadeh. "Automatic recognition of five types of white blood cells in peripheral blood." *Computerized Medical Imaging and Graphics* 35, no. 4 (2011): 333–343.

[11] Ramesh, Nisha, Bryan Dangott, Mohammed E. Salama, and Tolga Tasdizen. "Isolation and two-step classification of normal white blood cells in peripheral blood smears." *Journal of Pathology Informatics* 3 (2012).

[12] Putzu, Lorenzo, Giovanni Caocci, and Cecilia Di Ruberto. "Leucocyte classification for leukaemia detection using image processing techniques." *Artificial Intelligence in Medicine* 62, no. 3 (2014): 179–191.

[13] Nazlibilek, Sedat, Deniz Karacor, Tuncay Ercan, Murat Husnu Sazli, Osman Kalender, and Yavuz Ege. "Automatic segmentation, counting, size determination and classification of white blood cells." *Measurement* 55 (2014): 58–65.

[14] Mathur, Atin, Ardhendu S. Tripathi, and Manohar Kuse. "Scalable system for classification of white blood cells from Leishman stained blood stain images." *Journal of Pathology Informatics* 4, Suppl (2013).

[15] Ghosh, Pramit, Debotosh Bhattacharjee, and Mita Nasipuri. "Blood smear analyzer for white blood cell counting: A hybrid microscopic image analyzing technique." *Applied Soft Computing* 46 (2016): 629–638.

[16] Su, Mu-Chun, Chun-Yen Cheng, and Pa-Chun Wang. "A neural-network-based approach to white blood cell classification." *The Scientific World Journal* 2014 (2014): 796371.

[17] Adjouadi, Malek, Nuannuan Zong, and Melvin Ayala. "Multidimensional pattern recognition and classification of white blood cells using support vector machines." *Particle and Particle Systems Characterization* 22, no. 2 (2005): 107–118.

[18] Ghosh, Madhumala, Devkumar Das, Subhodip Mandal, Chandan Chakraborty, Mallika Pala, Ashok K. Maity, Surjya K. Pal, and Ajoy K. Ray. "Statistical pattern analysis of white blood cell nuclei morphometry." In *2010 IEEE Students Technology Symposium (TechSym)*, pp. 59–66. IEEE, 2010.

[19] Habibzadeh, Mehdi, Adam Krzyżak, and Thomas Fevens. "Comparative study of shape, intensity and texture features and support vector machine for white blood cell classification." *Journal of Theoretical and Applied Computer Science* 7, no. 1 (2013): 20–35.

[20] Schneider, Bendix, Geert Vanmeerbeeck, Richard Stahl, Liesbet Lagae, Joni Dambre, and Peter Bienstman. "Neural network for blood cell classification in a holographic microscopy system." In *2015 17th International Conference on Transparent Optical Networks (ICTON)*, pp. 1–4. IEEE, 2015.

[21] Ravikumar, S. "Image segmentation and classification of white blood cells with the extreme learning machine and the fast relevance vector machine." *Artificial Cells, Nanomedicine, and Biotechnology* 44, no. 3 (2016): 985–989.

[22] Zhao, Jianwei, Minshu Zhang, Zhenghua Zhou, Jianjun Chu, and Feilong Cao. "Automatic detection and classification of leukocytes using convolutional neural networks." *Medical and Biological Engineering and Computing* 55, no. 8 (2017): 1287–1301.

[23] Al-Dulaimi, Khamael, Vinod Chandran, Jasmine Banks, Inmaculada Tomeo-Reyes, and Kien Nguyen. "Classification of white blood cells using bispectral invariant features of nuclei shape." In *2018 Digital Image Computing: Techniques and Applications (DICTA)*, pp. 1–8. IEEE, 2018.

[24] Al-Dulaimi, Khamael, Kien Nguyen, Jasmine Banks, Vinod Chandran, and Inmaculada Tomeo-Reyes. "Classification of white blood cells using l-moments invariant features of nuclei shape." In *2018 International Conference on Image and Vision Computing New Zealand (IVCNZ)*, pp. 1–6. IEEE, 2018.

[25] Ko, B. C., J. W. Gim, and J. Y. Nam. "Cell image classification based on ensemble features and random forest." *Electronics Letters* 47, no. 11 (2011): 638–639.

[26] Ravikumar, S., and A. Shanmugam. "WBC image segmentation and classification using RVM." *Applied Mathematical Sciences* 8, no. 45 (2014): 2227–2237.

[27] Nassar, Mariam, Minh Doan, Andrew Filby, Olaf Wolkenhauer, Darin K. Fogg, Justyna Piasecka, Catherine A. Thornton et al. "Label-free identification of white blood cells using machine learning." *Cytometry Part A* 95, no. 8 (2019): 836–842.

[28] JS, Dr. "Gui based performance comparison of WBC segmentation and its classification." *Medical Hypotheses* 135 (2020): 109472.

[29] Vogado, Luis HS, Rodrigo de MS Veras, Alan R. Andrade, Romuere RV e Silva, Flavio HD De Araujo, and Fatima NS De Medeiros. "Unsupervised leukemia cells segmentation based on multi-space color channels." In *2016 IEEE International Symposium on Multimedia (ISM)*, pp. 451–456. IEEE, 2016.

[30] Kumar, P. S., and S. Vasuki. "Automated diagnosis of acute lymphocytic leukemia and acute myeloid leukemia using multi-SV." *Journal of Biomedical Imaging and Bioengineering* 1, no. 1 (2017): 1–5.

[31] Nasir, A. S. Abdul, M. Y. Mashor, and H. Rosline. "Unsupervised colour segmentation of white blood cell for acute leukaemia images." In *2011 IEEE International Conference on Imaging Systems and Techniques*, pp. 142–145. IEEE, 2011.

[32] Sajjad, Muhammad, Siraj Khan, Muhammad Shoaib, Hazrat Ali, Zahoor Jan, Khan Muhammad, and Irfan Mehmood. "Computer aided system for leukocytes classification and segmentation in blood smear images." In *2016 International Conference on Frontiers of Information Technology (FIT)*, pp. 99–104. IEEE, 2016.

[33] Kutlu, Hüseyin, Engin Avci, and Fatih Özyurt. "White blood cells detection and classification based on regional convolutional neural networks." *Medical Hypotheses* 135 (2020): 109472.

[34] Banik, Partha Pratim, Rappy Saha, and Ki-Doo Kim. "An automatic nucleus segmentation and CNN model based classification method of white blood cell." *Expert Systems with Applications* 149 (2020): 113211.

[35] Nazlibilek, Sedat, Deniz Karacor, Tuncay Ercan, Murat Husnu Sazli, Osman Kalender, and Yavuz Ege. "Automatic segmentation, counting, size determination and classification of white blood cells." *Measurement* 55 (2014): 58–65.

[36] Vincent, Ivan, Ki-Ryong Kwon, Suk-Hwan Lee, and Kwang-Seok Moon. "Acute lymphoid leukemia classification using two-step neural network classifier." In *2015 21st Korea-Japan Joint Workshop on Frontiers of Computer Vision (FCV)*, pp. 1–4. IEEE, 2015.

[37] Imran Razzak, Muhammad, and Saeeda Naz. "Microscopic blood smear segmentation and classification using deep contour aware CNN and extreme machine learning." In *Proceedings of the IEEE Conference on Computer Vision and Pattern Recognition Workshops*, pp. 49–55. IEEE, 2017.

[38] Yu, Wei, Jing Chang, Cheng Yang, Limin Zhang, Han Shen, Yongquan Xia, and Jin Sha. "Automatic classification of leukocytes using deep neural network." In *2017 IEEE 12th International Conference on ASIC (ASICON)*, pp. 1041–1044. IEEE, 2017.

[39] Tran, Thanh, Oh-Heum Kwon, Ki-Ryong Kwon, Suk-Hwan Lee, and Kyung-Won Kang. "Blood cell images segmentation using deep learning semantic segmentation." In *2018 IEEE International Conference on Electronics and Communication Engineering (ICECE)*, pp. 13–16. IEEE, 2018.

[40] Fan, Haoyi, Fengbin Zhang, Liang Xi, Zuoyong Li, Guanghai Liu, and Yong Xu. "LeukocyteMask: An automated localization and segmentation method for leukocyte in blood smear images using deep neural networks." *Journal of Biophotonics* 12, no. 7 (2019): e201800488.

[41] Liang, Gaobo, Huichao Hong, Weifang Xie, and Lixin Zheng. "Combining convolutional neural network with recursive neural network for blood cell image classification." *IEEE Access* 6 (2018): 36188–36197.

[42] Bani-Hani, Dana, Naseem Khan, Fatimah Alsultan, Shreya Karanjkar, and Nagen Nagarur. "Classification of leucocytes using convolutional neural network optimized through genetic algorithm." In *Proceedings of the 7th Annual World Conference of the Society for Industrial and Systems Engineering*. doi: 10.1371/journal. pone.0189259 2018.

[43] Hegde, Roopa B., Keerthana Prasad, Harishchandra Hebbar, and Brij Mohan Kumar Singh. "Feature extraction using traditional image processing and convolutional neural network methods to classify white blood cells: A study." *Australasian Physical and Engineering Sciences in Medicine* 42, no. 2 (2019): 627–638.

[44] Sharma, Mayank, Aishwarya Bhave, and Rekh Ram Janghel. "White blood cell classification using convolutional neural network." In *Soft Computing and Signal Processing*, pp. 135–143. Springer, Singapore, 2019.

[45] Wang, Justin L., Anthony Y. Li, Michelle Huang, Ali K. Ibrahim, Hanqi Zhuang, and Ali Muhamed Ali. "Classification of white blood cells with patternnet-fused ensemble of convolutional neural networks (pecnn)." In *2018 IEEE International Symposium on Signal Processing and Information Technology (ISSPIT)*, pp. 325–330. IEEE, 2018.

[46] H. Mohamed, Ensaf, Wessam H. El-Behaidy, Ghada Khoriba, and Jie Li. "Improved white blood cells classification based on pre-trained deep learning models." *Journal of Communications Software and Systems* 16, no. 1 (2020): 37–45.

[47] Choi, Jin Woo, Yunseo Ku, Byeong Wook Yoo, Jung-Ah Kim, Dong Soon Lee, Young Jun Chai, Hyoun-Joong Kong, and Hee Chan Kim. "White blood cell differential count of maturation stages in bone marrow smear using dual-stage convolutional neural networks." *PloS One* 12, no. 12 (2017): e0189259.

[48] Rehman, Amjad, Naveed Abbas, Tanzila Saba, Syed Ijaz ur Rahman, Zahid Mehmood, and Hoshang Kolivand. "Classification of acute lymphoblastic leukemia using deep learning." *Microscopy Research and Technique* 81, no. 11 (2018): 1310–1317.

[49] Alom, Md Zahangir, Chris Yakopcic, Tarek M. Taha, and Vijayan K. Asari. "Microscopic blood cell classification using inception recurrent residual convolutional neural networks." In *NAECON 2018-IEEE National Aerospace and Electronics Conference*, pp. 222–227. IEEE, 2018.

[50] Wang, Justin L., Anthony Y. Li, Michelle Huang, Ali K. Ibrahim, Hanqi Zhuang, and Ali Muhamed Ali. "Classification of white blood cells with patternnet-fused ensemble of convolutional neural networks (pecnn)." In *2018 IEEE International Symposium on Signal Processing and Information Technology (ISSPIT)*, pp. 325–330. IEEE, 2018.

[51] Wang, Qiwei, Shusheng Bi, Minglei Sun, Yuliang Wang, Di Wang, and Shaobao Yang. "Deep learning approach to peripheral leukocyte recognition." *PloS One* 14, no. 6 (2019): e0218808.

[52] Jha, Krishna Kumar, and Himadri Sekhar Dutta. "Mutual information based hybrid model and deep learning for acute lymphocytic leukemia detection in single cell blood smear images." *Computer Methods and Programs in Biomedicine* 179 (2019): 104987.

[53] Pandey, Prashant, Vinay Kyatham, Deepak Mishra, and Tathagato Rai Dastidar. "Target-independent domain adaptation for WBC classification using generative latent search." *IEEE Transactions on Medical Imaging* 39, no. 12 (2020): 3979–3991.

[54] Basnet, Jaya, Abeer Alsadoon, P. W. C. Prasad, Sarmad Al Aloussi, and Omar Hisham Alsadoon. "A novel solution of using deep learning for white blood cells classifi cation: Enhanced loss function with regularization and weighted loss (ELFRWL)." *Neural Processing Letters* 52, no. 2 (2020): 1517–1553.

[55] Patil, A. M., M. D. Patil, and G. K. Birajdar. "White blood cells image classification using deep learning with canonical correlation analysis." *IRBM* 42, no. 5 (2020): 378–389.

[56] Reena, M. Roy, and P. M. Ameer. "Localization and recognition of leukocytes in peripheral blood: A deep learning approach." *Computers in Biology and Medicine* 126 (2020): 104034.

4 Comparison and Performance Evaluation Using Convolution Neural Network-Based Deep Learning Models for Skin Cancer Image Classification

Snehal K. Joshi
Dolat-Usha Institute of Applied Sciences, Veer Narmad
South Gujarat University, Gujarat, India

4.1 INTRODUCTION

Studies carried out at different fronts and reports of the World Health Organization (WHO) have revealed that ultraviolet rays (UVR) are one of the most important factors primarily responsible for the development of melanoma.[1] Apart from preventive measures recommended in the guidelines to reduce the possibility of melanoma, an important factor is early detection, which can reduce serious threats. According to one study,[2] the number of melanoma patients in Europe ranges from 4 to 14 per million.

Skin texture and complexion play a major role in analyzing skin images. Skin texture analysis is important in identification and analysis as well as segmentation of the similarity and dissimilarity between specific areas. This helps to analyze features of the images obtained, feature extractions, and grouping according to similar features.

Skin pigmentation is a fundamental cause of various skin colors and shades. Pigmentation depends on geographical region, sunlight conditions in those particular areas, and specific sun rays that generate effects in the skin.

The term melanoma refers to a serious type of skin cancer. It is produced in the cells responsible for giving skin color – the pigment. Melanin is responsible for pigmentation. Particular cells called melanocytes produce melanin. There are various causes of deformation of melanocyte cells.[3] One reason is skin non-exposure to ultraviolet (UV) light. Lack of skin exposure to UV light raises the probability of

DOI: 10.1201/9781003230540-4

melanoma in humans. It is also seen that the risk factor of having melanoma is higher among those under the age of 40, particularly in European countries where sunlight exposure is a problem.[4] Although melanoma is fatal if not treated in the early stages, it is highly curable if detected early.

Four symptoms give visible signs of the possible development of melanoma. A mole with an unusual shape, having irregular borders separating the mole from surrounding skin and with a large shape, usually greater than 0.6 cm, are possible signs of melanoma. Melanoma is not only caused by the development of moles.[4]

A neoplasm is an unusual growth of cells that form a tumor or cyst. According to the WHO, by the end of 2020 approximately 1.81 million new cases of neoplasm were expected. Cases of neoplasm are growing linearly every year and threatening the health of the population.[5] However, not all neoplasms are fatal; some may be benign. Hence, early identification and classification of a neoplasm as malignant or benign is important.

There are many ways in which benign and malignant images are visibly different. However, three main aspects are seen:

1. The malignant condition is darker in color compared to the benign condition.
2. The area of the benign condition is generally less than 0.25 cm, whereas malignant conditions are larger than 0.25 cm.
3. The outer border of the benign condition that separates it from the skin is lighter, whereas in malignant conditions the border is darker.

In addition to these three major differences many other aspects can classify benign and malignant images.

Figure 4.1 shows images of malignant and benign conditions from a dataset. Various approaches are used for early detection of melanoma, including early

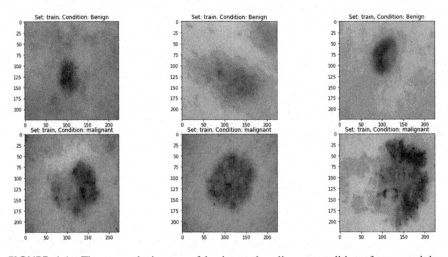

FIGURE 4.1   Three sample images of benign and malignant conditions from a training dataset.

detection through feature extraction of image and obtaining a dataset containing features, segmentation of images and extracting region of interest, use of pathological testing results, and biopsy results from. One way to predict melanoma and classify it with greater accuracy is based on a deep learning technique using computer vision. Convolutional neural network (CNN)-based classifier models can be used for early and accurate detection of malignant skin cancer. The present study relates to three CNN-based three. The study approach was to use the customized convolutional model and tune it to obtain best possible accuracy followed application of ResNet50 and VGG16-based models. The esults obtained were compared and discussed in terms of various parameters to signify performance matrix and adoption of a model that can be used for early detection of melanoma type for a given image. Of the three models applied, the first CNN-based classifier was based on two convolutional layers, one flattened layer followed by an input layer, and finally two hidden layers having an output layer. This basic model was trained and tested for different epochs and changing parameters. Another two CNN-based classifiers used were excellent image vision model architecture ResNet50 and VGG16. ResNet50 uses 50 layers. In comparison, VGG16 is an efficient architecture with approximately 138 million parameters. Both of these architecture-based models were used to classify a training dataset containing 2,637 benign and malignant types. The classifier model was tested using a dataset containing 660 images which were clinically verified. The dataset used processed skin cancer images available through an open dataset obtained from the International Skin Imaging Collaboration (ISIC), sourced from Kaggle.

In this study, three model performances were compared. Important questions to address were: (1) Does the stack of convolutional layers improve the performance matrix of the classifier model? (2) How do over-fitting problems arise due to a greater number of layers? (3) How are gradient descent, over-fitting, and under-fitting problems handled by the different models?

4.2 LITERATURE REVIEW

Many studies have been carried out on different approaches to the problem of classification when identifying the type of melanoma. Some studies apply approaches based on image processing and enhancement of images followed by feature extraction, region-of-interest extraction, and segmentation approaches.[6] Certain studies have been based on image feature extraction and their properties. Image analysis was also carried out using machine learning approach by applying various classifying algorithms that work on a dataset containing image features and related attributes like age, region, gender, etc.[7] Other studies have been based on neural network and self-extraction of features by the model by enhancing the images.[8] One important aspect of the approach in this study wass the CNN-based model. The CNN model is based on a deep learning approach that is trained using pathological identified and tested images; the same model is subsequently used to test unknown images to identify the model's performance.

Inlllal work was carried out by Esteva et al.[9] using a pre-trained classifier model-based CNN on GoogleNet Inception V3. Their model used a large dataset combination of 3,374 dermatoscopic images, including 129,450 clinically verified skin cancer images and yielded accuracy of 72.1% with error range of 0.9%. Another study carried out by Yu et al. [10] based on CNN with more than 50 layers using the ISBI dataset classified malignant and benign images with 85.5% accuracy. Following this, significant results were obtained by Haenssle et al. [11] using a CNN-based model to classify dermatoscopic images with sensitivity of 86.6%. Dorj et al. [12] proposed a model based on multiclass classification that used an approach based on pre-trained AlexNet Deep Learning CNN model and achieved 95.1% accuracy. Several more significant studies were carried out on this baseline; one, by Han et al., [13] obtained 96.00% accuracy with a variation margin of 1% using a CNN-based classifier.

In recent years important research has been carried out and researchers have applied different approaches. Hosny et al.[5] presented a technique based on deep learning that used a transfer learning approach. This study fine-tuned parameters and used data augmentation. The dataset is trained over Alexnet by application of the transfer learning approach. It applied the softmax activation function to the last layer of the AlexNet. The study measured the performance matrix measures of precision, specificity, sensitivity, and accuracy: the results showed 98.61% accuracy and 97.73% precision. Sensitivity and specificity were 98.33% and 98.93% respectively. Adekanmi and Serestina [6] proposed a model to detect melanoma lesions. Further, segmentation was based on an open-source dataset available through ISIC of dermoscopic images containing 2,000 images of skin lesions for the purpose of training. A trained model tested using 200 skin lesion images achieved 95% and 92% accuracy and dice coefficient respectively. Zhou et al.[7] proposed a training model based on an unsupervised spike timing-dependent plasticity learning rule. This model also comprises applied efficient temporal coding and a winner-takes-all (WTA) mechanism collectively that achieves 83.8% accuracy. Gurung and Gao[8] proposed a model that covers three aspects, including augmentation, boundary extraction, and feature extraction. This approach used exclusive OR with regression technique. As an outcome of this approach, search space is reduced, yielding better accuracy which is significantly higher by 1.2% compared to other models tested on databases such as PH2 and ISBI.

4.3 OBJECTIVES

The core objective of the study was to obtain a model that can classify a given skin image as benign or malignant. Various approaches can be applied for this purpose. However, the current study derived a CNN-based model for classification. The central idea of this study was based on the following objectives:

1. To classify a given skin image as benign or malignant with the greatest possible accuracy.
2. To enhance the model by tuning parameters to obtain greater accuracy.
3. To compare the model performance with results obtained by applying ResNET50 and VGG16-based classifier models.
4. To analyze model performance on a scale of performance matrix, including precision, recall, specificity sensitivity, and accuracy.

Different models based on the three approaches were used and based on their performance the models were enhanced. VGG16 and ResNet50 are widely proven and adopted architectures. Performance tuning and enhancement were core objectives. In addition, customized models were analyzed based on performance and the possibility of their adoption over proven architectures like ResNet50 and VGG16, particularly when a dataset is small in size.

4.4   METHODS AND METHODOLOGY

The dataset used for this classification was an open data source available from ISIC. It is a quality dermoscopic image of skin lesions available in jpg format with the dimension of 224 × 224 pixels.

There were 3,297 images available for this work, from which 1,800 images were classified after clinical verification as benign and 1,497 images as malignant. Dataset images were divided into training and testing sets. The division was in a random order resulting in two folders consisting of 1,440 malignant images in the training folder and 1,197 benign images in the testing folder. All images were jpg formats with dimensions of 224 × 224. The testing folder contained 360 benign and 300 malignant jpg images of dimension 224 × 224.

Since the dataset wass not large, the aim was that the classifier should fit the need for trainingwith the best possible performance measures. The adoption of a classification model had the objectives of the study as the prime focus. There are two architectures, ResNet50 and VGG50, which have been proven to be highly effective in building an image classification model. Apart from these two models, we also aimed to customize and implement a CNN model.

Based on the following methodology, different classifiers were implemented on the dataset. A CNN-based basic model was used with different variations of model. This was followed by implementation of two prominent architectures, VGG16 and ResNet50. As depicted in Table 4.1, comparison was made of all seven classifier architectures.

Classifier model 1: This model architecture was based on dimensionality reduction of images to 224×224 dimensions for training and testing followed by augmentation using zoom value 0.3 and reshaped to a scale of 1/0.255 using vertical flip. The model is explained as shown in Figure 4.2.

Classifier model 2: This had variations compared to model 1, including reduction in the dimensions of training images. It used 150×150 image dimensions. The aim of this model was to analyze the effect of dimension reduction on model performance.

Classifier model 3: This model was an extension of model 1. It was selected after evaluation of model 1 and model 2 performance. Epochs were changed from the selected model 1 or model 2.

Classifier model 4: An additional convolution block consisting of two layers followed by a pooling layer that was included in the modelc3-based architecture was modified and trained. All three blocks contained two layers, each consisting of 16, 32, and 64 filters respectively.

TABLE 4.1
Architecture and parameter comparison of seven implemented models

Parameters	Model 1	Model 2	Model 3	Model 4	Model 5	Model 6	Model 7
Model type	Customized	Customized	Customized	Customized	Customized	VGG16	ResNet50
Epochs	15	15	25	25	25	25	25, 50 and 100
Batch size	64	64	64	64	64	32	64
Dimension	224×224	150×150	224×224	224×224	224×224	224×224	224×224
Nos. of convolution blocks	02 with 02 layers	02 with 02 layers	02 with 02 layers	03 with 02 layers	03 with 02 layers	04 with total 13 layers	Pre-trained model weights with 50 layers including input and output layers.
Filters	16 and 32	16 and 32	16 and 32	16,32,64	16,32,64	64,128, 256, 512,512	
Pooling	Maxpool	Maxpool	Maxpool	Maxpool	Maxpool	Maxpool	
Hidden layers	02	02	02	02	03	02 with 1,028 nodes	
Activation function	ReLu	ReLu	ReLu	ReLu	ReLu	ReLu	ReLu
Optimizer	Adam	Adam	Adam	Adam	Adam	Adam	Adam
Output layer activation function	Sigmoid	Sigmoid	Sigmoid	Sigmoid	Sigmoid	Sigmoid	Softmax

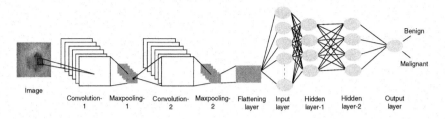

FIGURE 4.2 Convolutional neural network architecture implemented with three different variations.

Classifier model 5: By adding one extra hidden layer in a dense layer, model 4 was modified and trained for 25 epochs. This model contained three blocks of convolution layers, each consisting of two layers and three hidden layers.

Classifier model 6: VGG16 was a highly efficient and performance-driven architecture with 16 layers, collectively including approximately 138 million parameters. The approach of VGG16 architecture was its unique focus, having consistency in a convolution layer of 3×3 filters with a single stride. It used a max-pool layer of filter size 2×2 with two strides and similar padding. The VGG16 architecture used a flattening layer and two fully connected layers and finally one output layer, with one node as output node that used the softmax function.

Classifier model 7: Using the pre-trained ResNet50, the model was trained using the training dataset images. The model was modified at a convolutional layer. Two changes were made and the model was customized. It had three layers of kernel size 7×7 with a stack of 64 kernels and a stride of two, followed by maxpooling with stride of two, and finally a convolutional layer having 1×1 sizes of 64 kernels, 3×3 size of 64 kernels, and 1×1 size of 256 layer.

All models were trained over the graphics processing unit (GPU) available through Google's colab Jupiter notebook. For the purpose of performance measurement of all seven models, a performance matrix analysis used eight performance matrices – model accuracy, precision, recall, specificity, sensitivity, F1-score, false discovery rate (FDR) and false-negative rate (FNR) measures. Performance of all seven models was based on model architecture parameters, including filter size, number of convolution layers, number of dense layers, activation function used, optimizer, number of dense layers, and loss function.

Apart from the architecture of the model, learning rate and early stopping parameters were also analyzed to evaluate performance. The performance of the model also depended on the number or epochs, batch size, and step per epochs. Accuracy of testing and training as well as loss were compared to assess model performance. Validation set accuracy was also obtained to check how well the model performs on a validation image set. An important aspect of analyzing model perform ance is based on the confusion matrix. The confusion matrix was obtained to measure recall and precision; sensitivity and specificity were also obtained.

4.5   DATA ANALYSIS AND PERFORMANCE EVALUATION

CNN models based on different architects are compiled and their performances were evaluated. All models were trained over a limited training dataset consisting of 2,547 training images, including 1,440 benign and 1,197 malignant clinically veri-fied skin images of size 224×224 dimensions. The testing dataset consisted of 660 skin images: 360 benign images and 300 malignant images of uniform dimension 224×224.

4.5.1   PERFORMANCE EVALUATION OF MODEL 1

Model 1 consisted of two convolution layers. The first block of the convolutional layer had 16 filters, kernel size was 3×3, activation function was ReLu. The second layer had 16 filters.

The first training epoch took 22 seconds for each of 41 steps, carrying an elapse time of 537 ms per step. Training loss and accuracy for the first epoch was observed as 0.3391 and 0.8395 respectively with validation loss and accuracy of 0.4120 and 0.7953 respectively. A comparison of loss and accuracy for validation and training is depicted in Figure 4.3. A confusion matrix was obtained for a validation dataset. The confusion matrix is:

$$
\begin{aligned}
TP &= 309 \quad FP = 51 \\
FN &= 53 \quad\; TN = 247
\end{aligned}
\tag{4.1}
$$

The model performance shown in Figure 4.3 for training and validation has been presented graphically. The model performance matrix is depicted in Table 4.2.

4.5.2   PERFORMANCE EVALUATION OF MODEL 2

This second model was a variation on model 1; the only major change was in image dimension. Dataset images were reduced to 150×150. The model was trained for 15 epochs using the training set consisting of 2,637 images. A confusion matrix was

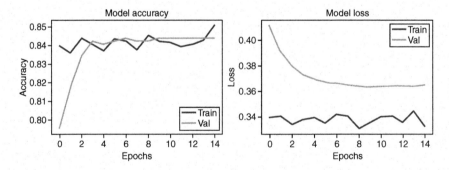

FIGURE 4.3   Model 1 accuracy and loss comparison for training and validation.

TABLE 4.2
Performance evaluation of model 1 based on different parameters

Accuracy	Precision	Recall	Specificity	Sensitivity	F1-score	FDR	FNR
82.24%	82.88%	82.33%	82.88%	85.35%	82.06%	17.12%	14.65%

FDR, false discovery rate; FNR, false-negative rate.

TABLE 4.3
Performance evaluation of model 2 based on different parameters

Accuracy	Precision	Recall	Specificity	Sensitivity	F1-score	FDR	FNR
81.81%	78.88%	86.58%	77.10%	86.58%	81.01%	22.90%	13.42%

FDR, false discovery rate; FNR, false-negative rate.

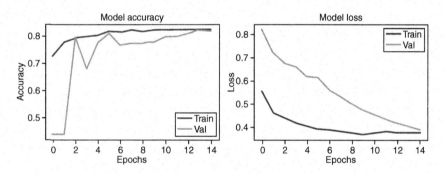

FIGURE 4.4 Model 2 accuracy and loss comparison for training and validation.

obtained for the testing dataset of size 660, consisting of 360 benign and 300 malignant images. It was observed that true positive (TP) = 284, false positive (FP) = 76, false negative (FN) = 44 and true negative (TN) = 256 for 600 validation images. Here type 1 error was 76 and type 2 error was 44. Training and validation performance can be seen in Figure 4.4. Model performance was measured on eight different parameters (Table 4.3).

4.5.3 PERFORMANCE EVALUATION OF MODEL 3

This third model was customized. Its convolution layer consisted of two layers. The model was trained for 23 epochs using a training set consisting of 2,637 images. The input images were of 224×224 dimension and had three channels. The architecture of the model is depicted in Table 4.1. The confusion matrix is given below and a performance evaluation of eight parameters is shown in Table 4.4.

$$\text{Confusion Matrix:} \begin{bmatrix} \begin{bmatrix} 262 & 98 \end{bmatrix} \\ \begin{bmatrix} 30 & 270 \end{bmatrix} \end{bmatrix} \tag{4.2}$$

Model 3 observations depict TP = 262, FP = 98, FN = 30, and TN = 270 for the testing set consisting of 660 images for validation. Here type 1 error is 98 and type 2 error is 30. The performance analysis of model 3 for 25 epochs can be seen in Figure 4.5. Validation accuracy was 80.60%, precision or positive predictive value was 72.77%, recall was 89.72%, specificity 73.36%, and sensitivity 89.72%. The F1-score is the harmonic mean of precision over recall; and for model 3 the F1-score was 80.83%. FDR is related to type 1 error, and for model 3 FDR was 26.64%. FNR addresses type 2 error, and for model 3 FNR was 10.28%. A summary of the performance matrix for model 3 based on these eight parameters is given in Table 4.4.

4.5.4 PERFORMANCE EVALUATION OF MODEL 4

Variation in the fourth model compared to model 3 was the inclusion of one additional convolution block containing 16 filters of size 3×3, 32 filters of size 3×3, and 64 filters of size 3×3 at convolution block 1, block 2, and block 3 respectively. Each block contained two convolution layers followed by a pooling layer (Maxpool) of pool size 2×2. All other parameters and filters were kept similar to model 3. The

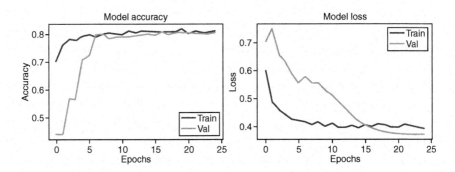

FIGURE 4.5 Model 4 accuracy and loss comparison for training and validation.

TABLE 4.4
Performance evaluation of model 3 based on different parameters

Accuracy	Precision	Recall	Specificity	Sensitivity	F1-score	FDR	FNR
80.60%	72.77%	89.72%	73.36%	89.72%	80.83%	26.64%	10.28%

FDR, false discovery rate; FNR, false-negative rate.

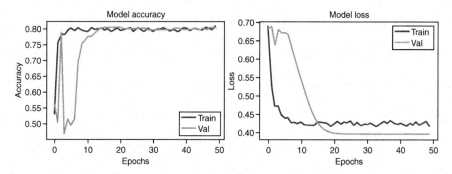

FIGURE 4.6   Model 4 accuracy and loss comparison for training and validation.

TABLE 4.5
Performance evaluation of model 4 based on different parameters

Accuracy	Precision	Recall	Specificity	Sensitivity	F1-score	FDR	FNR
82.27%	75.20%	90.90%	75.20%	90.90%	82.35%	24.80%	9.10%

FDR, false discovery rate; FNR, false-negative rate.

odel was trained for 25 epochs. It contains 12,898,721 total parameters, of which 12,898,593 are trainable parameters. The model was trained for 25 epochs and epoch-wise training accuracy and loss as well as validation accuracy and loss were obtained. The observations are depicted in Figure 4.6. Model 4 confusion matrix observations were obtained for 660 images for validation. Type 1 error was 90 and type 2 error was 27.

$$\text{Confusion Matrix}: \begin{bmatrix} [270 \ 90] \\ [27 \ 273] \end{bmatrix} \tag{4.3}$$

Based on the confusion matrix, performance measurements for model 4 considering eight different parameters were obtained and are depicted in Table 4.5.

4.5.5   PERFORMANCE EVALUATION OF MODEL 5

No significant changes were made in the architecture of this model except for the addition of one extra hidden layer modifying model 4. All other parameters and features of the model, all filters, and convolutional layers, pooling layer, and activation function were similar to model 4. The model contained 3,267,267 parameters, out of which 3,267,489 were trainable parameters. The model was trained for 50 epochs and training accuracy and loss, validation accuracy and loss were obtained as shown in Figure 4.6.

$$\text{Confusion Matrix} : \begin{bmatrix} [250 & 110] \\ [19 & 281]] \end{bmatrix} \qquad (4.4)$$

Type 1 error was 110 and type 2 error was 19. Performance measurements for model 5 considering eight different parameters were obtained and are summarized in Table 4.6.

4.5.6 PERFORMANCE EVALUATION OF MODEL 6 – VGG16

The VGG16 model was implemented on image training and testing dataset of dimensions 224×224 for 25 epochs and batch size of 32. Initially the training image set, dimension 224×224, was rescaled for zoom factor 0.3 and vertical flip. The model consisted of five blocks collectively having 16 layers in a sequential manner.

The first epoch took 52 seconds for 82 steps consisting of batch size 32. The step execution average timing for the first epoch was 625 ms/step. The first epoch gave training accuracy of 0.5457 and validation accuracy of 0.5625. The final 25th epoch took 51 seconds of execution time for 82 steps consisting of batch size 32 images with an average of 617 ms/step. The first epoch gave training accuracy of 0.5455 and validation accuracy of 0.5625 (Figure 4.7). The VGG16 model did not show any significant improvement in training or validation accuracy even at the end of the 25th epoch. In contrast, VGG16 model validation accuracy remained flat without any

---

TABLE 4.6
Performance evaluation of model 5 based on different parameters

Accuracy	Precision	Recall	Specificity	Sensitivity	F1-score	FDR	FNR
80.45%	69.44%	92.93%	71.86%	92.93%	81.33%	28.14%	7.07%

FDR, false discovery rate; FNR, false-negative rate.

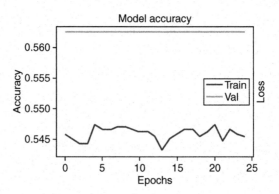

FIGURE 4.7   Accuracy comparison for training and validation for VGG16 model.

improvement, with only 56.25% accuracy. The model was trained using Adam optimizer and binary_crossentropy loss function. It contained 139,506,497 parameters, of which 139,506,497 are trainable parameters.

4.5.7 Performance Evaluation of Model 7 – ResNet50

In the ResNet50 model pre-trained weights were used to validate the validation set. The first weight file was used to test the validation dataset. The second training file's initial weights were used for kernel transfer learning. The ResNet50-based model used training and validation image sets of image size 224×224 using transfer learning and tested for ten epochs with batch-size of 16, pooling layer Maxpool, Optimizer 'Adam' with binary_crossentropy loss function. Since there were fewer epochs to obtain, it is possible that the pre-trained model resulted in an over-fitting problem and gave greater accuracy.

The model was executed for ten epochs for a batch size of 16. The first epoch, executed for 25 seconds with training loss and accuracy, obtained 0.2983 and 0.8620 respectively. Validation loss and accuracy were 0.0655 and 0.9730 respectively. From the very first epoch, accuracy was 97.30%. The tenth epoch improved performance with validation loss of 0.0617 and validation accuracy of 0.9790. The validation loss decreased at the tenth epoch compared to the first epoch from 0.0655 to 0.0565. Accuracy was also improved from 97.30% at the first epoch to 97.90% at the tenth epoch. When the epoch values were increased from the initial epoch value of ten to 25, 50, and 100, accuracy was 79.22%, 76.38%, and 72.72% for 25, 50, and 100 epochs, respectively. Hence, where there are fewer epochs in the pre-trained model, over-fitting is the major reason for greater accuracy.

4.6 RESULT ANALYSIS AND INTERPRETATION

The performance analysis of all five customized CNN-based models can be summarized as shown in Table 4.7. The greatest accuracy (82.27%) was obtained with model 4, but the highest precision (82.88%) and recall (92.93%) were obtained with model 1 and model 5 respectively. The highest specificity was obtained with model 1. F1-score was also significantly higher for model 1. Model 4 had an F1-score of 82.35%, which is the highest score among all models. FDR and FNR are vital measures that address the performance and capability of the model to handle type 1 and type 2 errors. FDR was lowest with model 1. The FNR was lowest at 7.07% for model 5. Model 1 performed significantly better in handling type 1 error but was not optimum in handling type 2 error. It addressed type 2 error significantly better compared to other models. As seen in Table 4.7, validation accuracy was comparatively higher (above 82%) for model 1 and model 4. Sensitivity was close to 90% for model 3, model 4, and model 5. FDR was lowest for model 1 and FNR was highest. At the other end, FNR was lowest for model 5 but FDR was highest. Considering the VGG16 and ResNet50-based models, VGG16 provided validation accuracy of 56.25%, which is significantly low compared to the customized CNN-based models that have much fewer convolution layers. In comparison, ResNet50 demerits the model performance when the number of epochs was increased to train the model.

TABLE 4.7

Performance comparison and summary of five customized models

Model	Accuracy	Precision	Recall	Specificity	Sensitivity	F1-score	FDR	FNR
Model 1	82.24%	82.88%	82.33%	82.88%	85.35%	82.06%	17.12%	14.65%
Model 2	81.81%	78.88%	86.58%	77.10%	86.58%	81.01%	22.90%	13.42%
Model 3	80.60%	72.77%	89.72%	73.36%	89.72%	80.83%	26.64%	10.28%
Model 4	82.27%	75.20%	90.90%	75.20%	90.90%	82.35%	24.80%	9.10%
Model 5	80.45%	69.44%	92.93%	71.86%	92.93%	81.33%	28.14%	7.07%

FDR, false discovery rate; FNR, false-negative rate.

Validation accuracy was 79.22% for 25 epochs and 76.38% for 50 epochs but significantly decreased to 72.72% when the number of epochs was increased to 100.

4.7 CONCLUSION

Based on the performance evaluation of all five models as summarized in Table 4.7 we can conclude important findings as follows:

1. All customized CNN-based models obtained accuracy results of greater than 80%.
2. A precision and recall tradeoff wass observed for all models. The F1-score for all models was higher than 80%, signifying a good harmonic mean between precision and recall.
3. Comparing the performance matrix of model 2 with that of model 1, accuracy, recall, and precision are not improved with a reduction in dimension.
4. Type 2 error is better addressed by model 5 compared to the other models. Sensitivity was 92.93% and FNR was 7.07%. The false negatives were better addressed by model 5 where three hidden layers were used with 256, 128, and 64 nodes. However, type 1 error was increased in this case compared to previous models.
5. Interesting results were observed with the VGG16 model. When ining the model was trained for 25 epochs, the validation accuracy showed a flat result and did not show any improvement, with 56.25% accuracy. Although there were 139,506,497 trainable parameters, the VGG16-based model did not show any significant classification results. The accuracy graph for 25 epochs showed how poorly the classifier performs (Figure 4.8).
6. Another interesting observation was made in the case of the ResNet50 classifier model. Training the model for ten epochs showed a significant improvement in accuracy from the first epoch: from 86.20% to 97.30% at the tenth epoch. Validation loss was decreased at the tenth epoch compared to the first epoch, from 0.0655 to 0.0565. But, when the epochs were increased epochs to

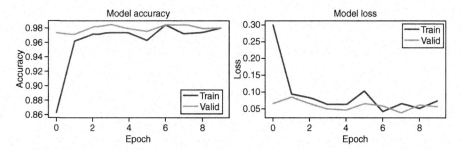

FIGURE 4.8   Model 5 accuracy and loss tradeoff for training and validation.

25, 50, and 100, the model performance decreased drastically. For 25, 50, and 100 epochs the model had 79.22%, 76.38%, and 72.72% accuracy, respectively. With increase in accuracy, validation accuracy decreased. Where there were fewer epochs in the pre-trained model based on ResNet50, over-fitting was the major reason for greater accuracy.

7. The customized model showed that increasing the convolutional layer with more filters does not affect accuracy and other performance measures. However, by adding a hidden layer, as seen in model 5 performance, type 2 error can be addressed significantly.

Customized CNN-based models performed significantly well compared to large convolutional models based on architectures like VGG16 and ResNet50, particularly when the training and validating image dataset was not very large. However, all models are trained by augmenting the training dataset; there is scope to add more augmentation parameters, which may affect the model performance. The learning rate can be changed and the model can be tuned by changing the learning rate and optimizers. There is great scope for improving the model but it is also evident that a large number of convolutional layers does not guarantee greater accuracy, specificity, and sensitivity. In such types of models addressing type 2 error is a major concern and of course not at the cost of type 1 error.

REFERENCES

[1] Rigel, D. "Cutaneous Ultra-Violet Exposure and its Relationship to the Development of Skin Cancer." *J Am Acad Dermatol* 2008; 58(5 suppl 2): S129–S132.

[2] Pašić, A., Lipozenčić J. i sur. Maligni melanom U. *Dermatovener-ologija*. Zagreb: Medicinska naklada; 2005, pp. 493–494.

[3] F.J. Lejeune (Editor-in-Chief), Storkus W.J., "Thirty years of therapeutic innovation in melanoma research." *Melanoma Res* 2021; 31(2): 105–107.

[4] Coleman W.B., Tsongalis G.J. (2018), *Molecular Pathology* (2nd ed.). Academic Press, pp. 71–97.

[5] Hosny K.M., Kaseem M.A., Foaud M.M. "Skin Cancer Classification using Deep Learning and Transfer Learning." Cairo International Biomedical Engineering Conference (CIBEC), 2018, pp. 90–93.

[6]   Adekanmi A. Adegun, Serestina V. "Deep Learning-Based System for Automatic Melanoma Detection." *Access IEEE* 2020; 8: 7160–7172.

[7]   Zhou Q., Shi Y., Xu Z., et al. "Classifying Melanoma Skin Lesions Using Convolutional Spiking Neural Networks with Unsupervised STDP Learning Rule." *Access IEEE* 2020; 8: 101309–101319.

[8]   Gurung S., Gao Y.R. "Classification of Melanoma (Skin Cancer) using Convolutional Neural Network." *2020 5th International Conference on Innovative Technologies in Intelligent Systems and Industrial Applications (CITISIA)* 2020, pp. 1–8.

[9]   Esteva A., Kuprel B., Novoa R.A., et al. "Dermatologist-Level Classification of Skin Cancer with Deep Neural Networks." *Nature* 2017; 542(7639): 115–118.

[10] Yu L., Chen H., Dou Q., Qin J., Heng, P.-A. "Automated Melanoma Recognition in Dermoscopy Images via Very Deep Residual Networks." *IEEE Trans Med Imaging* 2017; 994–1004.

[11] Haenssle H.A., Fink C, Schneiderbauer R, et al. "Man Against Machine: Diagnostic Performance of a Deep Learning Convolutional Neural Network for Dermoscopic Melanoma Recognition in Comparison to 58 Dermatologists." *Ann Oncol* 2018; 29(8): 1836–1842.

[12] Dorj U.O., Lee K.K., Choi J.Y., et al. "The Skin Cancer Classification Using Deep Convolutional Neural Network." *Multimed Tools Appl* 2018; 77: 9909–9924.

[13] Han S.S., Kim M.S., Lim W., et al., "Classification of the Clinical Images for Benign and Malignant Cutaneous Tumors using a Deep Learning Algorithm.", *J Investig Dermatol* 2018; 138(7): 1529–1538.

5 A Review of Breast Cancer Detection Using Deep Learning Techniques

Abhishek Das
ITER, Siksha 'O' Anusandhan, Bhubaneswar, Odisha, India

Mihir Narayan Mohanty
ITER, Siksha 'O' Anusandhan, Bhubaneswar, Odisha, India

5.1 INTRODUCTION

Cancer tissue has two types of cancer cells, known as maturable and non-maturable. The tissue is composed of maturable types with few cells of non-maturable type. It has both genetic and environmental causes. Most cancers are initiated by malignant tumors. These tumors have rapid growth. Breast cancer is one of them. Non-invasive types of cancer are within the milk ducts. In this case, there is no growth or spread into any surrounding tissue. Invasive ductal carcinoma is a common type, comprising 80% of all breast cancers. Breast cancer is notified as the highest-incidence disease among all types of cancer according to the report provided in [1] in 2020. A graphic representation of this study is provided in Figure 5.1. The main reasons for breast cancer include obesity, ultraviolet radiation, and infections. Symptoms of breast cancer include development of a lump in the breast or armpit, dimpling of breast skin, swelling or thickness in any part of the breast, nipple bleeding or pain, and abnormal change in breast shape or size [2]. Early detection of breast cancer after such symptoms have been observed can prevent its development and also death.

The procedure for cancer detection and classification involves imaging to visualize the location, shape, and size of the developing tissue causing cancer. Imaging techniques include mammography, computed tomography (CT), magnetic resonance imaging (MRI), histopathology, ultrasound (US), thermography, and microwave imaging.

Mammogram: Mammograms have been in use since 1960 for breast cancer screening. Restrictions to its use include age, family history, and tissue density [3]. The side effects of mammography include tissue damage for elderly patients. Young patients have dense breast tissue, so in most cases, a mammogram fails to detect the presence of any cancerous tissue.

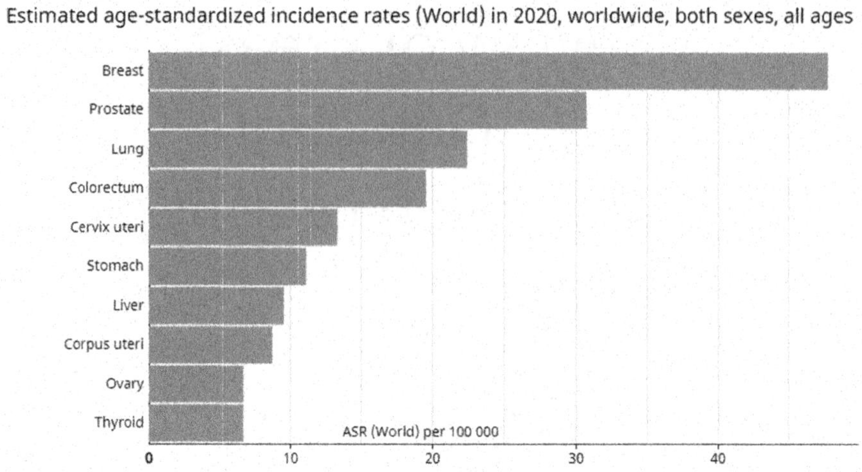

Estimated age-standardized incidence rates (World) in 2020, worldwide, both sexes, all ages

FIGURE 5.1   Estimated incidence rates of cancer in 2020 (worldwide). Adapted from [1].

Computed tomography: CT scanning results in two-dimensional slices of the breast taken from different angles using the X-ray technique. It involves a contrast injection in the arm to improve the imaging result with a high-quality reconstruction method. This injection may affect kidney functionality. Radiation is also harmful to patients [4].

Magnetic resonance imaging: MRI is another tool to detect the presence of unwanted growth inside the breast. It involves the application of radio waves in the presence of a magnetic field. Gadolinium is injected into the patient's blood to observe the condition of the blood vessels of the breast. MRI involves high cost and is less able to detect lower-grade breast cancer tissue [5].

Ultrasound: US is the safest option for breast cancer detection as it involves only the application of sound waves. However, it does have limitations, such as inability to detect cancer in early stages and high false-positive chances [6].

Histopathology: The microanatomy of breast cells and tissues can be studied for breast cancer detection using histopathology imaging. The method involves the use of microscopes and hematoxylin and eosin to stain the breast tissue, and it takes pathologists hours to observe the variations between healthy and cancerous tissues [7].

Thermography: Infrared thermography provides two-dimensional images depending upon temperature variations in the breast due to the presence of cancer. This method is free from any use of radiation but it results in high values of false-positive and false-negative detection [8].

Microwave imaging: The application of antennas in biomedical microwave imaging is an emerging field of research. Antenna-based microwave imaging has been proposed in [9] for breast cancer detection. Differentiation between cancerous and normal tissue was studied in their work, and also between the subtypes of breast cancers, i.e., benign or malignant.

Two techniques have been adopted for detection: complex natural response and microwave imaging using a Vivaldi antenna. This method results in three-dimensional (3D) images of the breast phantom. This imaging technique can be used by researchers in other application areas and using the images classification and detection can also be performed with deep learning-based techniques. In another work, the authors proposed using a planner array antenna arranged in a spherical pattern to obtain microwave imaging data to detect the tumor in the breast [10]. The methods follow a systematic use of antennae one at a time. One antenna is used as a microwave transmitter; at the same time other antennae are used as receivers and the pattern continues for all the participating antennae. The model was tested with metamaterial with the same dielectric properties as human breast over two different frequencies, i.e., 5 and 12 GHz. The results obtained from 12 GHz radiation were better in comparison to 5 GHz.

The application of an array of patch antennae has been proposed for breast cancer detection [11]. The antenna array is designed using low-profile aperture stacked patch antennae with an operating frequency range of 2.2–13.5 GHz. To remove the effects of reflections from skin an artifact removal scheme was adopted. The constructed image of the breast is then used for cancerous tumor detection.

Artificial neural networks (ANNs) can achieve human-like observation in the field of image processing and many other application areas. The difference between ANN and the human biological neural system is observed in synchronization. ANN functions concurrently whereas the human neural network works asynchronously.

Various machine learning techniques have been developed and used for breast cancer detection and classification. Methods like support vector machine (SVM), K-nearest neighbor, decision tree, and random forest are broadly applied in the detection of deadly breast cancer [12–15]. Recent developments in ANN in the form of deep learning mean that it has emerged as the preferred technique for detection of breast cancer from any kind of image datasets in single as well as hybrid system architectures [16–19].

The rest of this chapter is organized in the following manner. The second section describes state-of-the-art methods. This section includes medical aspects, traditional methods, machine learning approaches, and deep learning approaches. The third section provides a brief description of the proposed method used for breast cancer detection. The fourth section provides the results obtained from the proposed method and discussion. Finally, the fifth section concludes the chapter. It includes what is obtained, how it is used for better results and possible applicability for researchers, and future scope.

5.2 LITERATURE SURVEY

Medical imaging provides the current conditions of targeted organs; this is the reason why most medical centers are well equipped with such advanced technology-based scanning machinery. With the application of image processing, it is now possible to extract information from scanned images that are comparable to the diagnosis of experienced doctors. Therefore, researchers have been working to improve the performance based on this concept and developing methods for better diagnosis of disease. Various works have been developed in the field of breast cancer detection. In recent years methods of detection and classification have varied from the application of standard machine learning techniques to deep learning, from a single classifier-based model to an ensemble of different classifiers. The datasets have also varied from work to work. The types of datasets include mammography, CT, MRI, histopathology, US, thermography, and microwave imaging.

5.2.1 MACHINE LEARNING APPROACHES

A concatenated model containing three machine learning techniques has been proposed for breast cancer classification [20]. The important parameters of the data were identified and data size was reduced using principal component analysis. The resized data were then passed through a multiple-layer perceptron (MLP) model for feature generation. These features were then used for training and classification by the SVM model. The model was verified using the Manuel Gomes dataset. This dataset contains ten parameters: age of the participants, body mass index, glucose, insulin, adiponectin, leptin, chemokine monocyte chemoattractant protein, homeostasis model assessment, resistin, and the corresponding labels representing whether the patient has cancer or is healthy. The model provided 86.97% classification accuracy; this figure needs to be improved to compete with deep learning-based results.

Application of SVM and MLP for semantic segmentation and detection of breast cancer in histopathology images is found in [21]. The authors developed this machine learning algorithm to obtain nucleus-level annotations. The histopathology images were first preprocessed for color normalization and enhancement. Then texture features were extracted using gray level co-occurrence matrix, filter banks, and local binary patterns (LBP). These features were fed into SVM and MLP for training and classification. The model has been tested on the BreakHis and KMC datasets. Out of all combinations of features and classifiers, MLP trained with features obtained by the combination 'filter bank+ color' provided a F1-score of 0.83.

An online platform for breast cancer detection is proposed using the RCA algorithm, where R stands for refinement, C stands for correlation, and A stands for adaptive [22]. The three algorithms are used to fine-tune the features for classification by the AdaBoost-based tree model. This technique has been verified on the INbreast and CBIS datasets and provided accuracy values of 87.07% and 93.30% respectively. The results obtained by this method need to be improved to compete with deep learning-based methods.

A computer-aided breast cancer detection system has been proposed using data clustering and the Adaboost technique [23]. The method was designed to process the BUS ultrasound image dataset. The process involved feature extraction using suggestions given by doctors, and a feature-scoring scheme. Then a bi-clustering algorithm was applied to extract column-oriented patterns to create training data. Different rules were created to build the Adaboost-based classifier for the classification of breast ultrasound images into malignant or benign. Accuracy and sensitivity obtained in this method were 95.75% and 96.26% respectively. The handcrafted feature extraction makes this method complex in comparison to deep learning-based auto feature extraction methods.

Another work based on ultrasound image processing using texton filter banks and local binary pattern features has been proposed for breast cancer detection and classification into two subtypes, i.e., benign and malignant [24]. Textons are two-dimensional images formed by the microstructures of ultrasound images. Features generated by the use of Leung-Malik filter, Schmid filter, maximum response filter, and Adasyn synthetic sampling were then passed through a feature reduction technique using locality-sensitive discriminant analysis (LSDA). The method was verified on the dataset collected by the authors, which included 147 normal, 210 benign, and 91 malignant ultrasound images. This method provided 96.1% accuracy when the classifier was a probabilistic neural network.

Cancer detection in terms of mitotic cell count is another approach in breast cancer detection. For this purpose, authors have used a deep belief network (DBN) for mitotic and non-mitotic cell classification [25]. The DBN model was combined with a multi-class classifier developed by using neural network, non-linear SVM, linear discriminant, and decision trees. The method was verified on the Mitos dataset available at MITOS-ATYPIA 2014 contest and regional cancer center (RCC). The method provided an F-score of 84.29% and 75% for MITOS and RCC datasets respectively.

5.2.2 DEEP LEARNING APPROACHES

Stacked autoencoder (SAE)-based reconstruction has been used to improve the performance of breast cancer imaging by magnetic detection electrical impedance tomography [26]. Deep neural network (DNN-based SAE was proposed to reconstruct the image from the conductivity distribution trained with magnetic field data. The result obtained by this method has been projected to be more comparable to backpropagation and sensitivity algorithms in the presence of 30-dB noise with relative loss of 15.28%, 24.90%, and 137.19% respectively.

Cancer detection in terms of mitotic cell count is another approach in measuring the grade of breast cancer [27]. Various works have been developed using these imaging techniques with different detection and classification models. Mitosis count represents the degree of seriousness of breast cancer. It is challenging work when the size of mitosis is very small. This problem has been considered and focused on by authors [28], who have tried to count the large as well as small mitosis present in the breast cells from histopathology images. Deep learning-based techniques have been adopted for segmentation as well as for the detection of mitosis. Atrous fully connected convolutional neural network (A-FCNN) and multi-scale region-based convolutional neural network (MS-RCNN) models have been utilized for segmentation and detection respectively. Atrous CNN means dilated CNN has been used designed with dilation rates of 1, 2, and 3 to the second and third layers of CNN. Authors have used three publicly available datasets, ICPR 2012, ICPR 2014, and AMIDA 13, to validate the model and obtained F-score values of 0.902, 0.495, and 0.644 respectively. Since the deep learning-based models are capable of auto feature extraction, the use of the segmentation step makes the system complex.

The size of histopathology whole slide images varies from 40,000 × 40,000 to 100,000 × 100,000 pixels. Thus, instead of applying the whole image to the model, different patches containing useful information were extracted and then fed to the CNN model for training; this is termed multi-instance CNN [29]. A multi-instance pooling layer was introduced between the fully connected layers of the CNN model. The model is verified on the BreakHis, IUPHL, and UCSB datasets and the accuracy values obtained were 93.06%, 96.63%, and 95.58% respectively.

Mislabeled patches extracted from histopathology whole slide images have been considered for correction and further utilization in training for benign and malignant breast cancer classification by the authors [30]. Anomaly detection was performed by a generative adversarial network termed Ano-GAN and then features were extracted for training and classification by DenseNet-121. The proposed model in their work was verified on the BreakHis dataset with fivefold cross-validation; 99.13% accuracy and 99.38% F1-score were obtained.

A region-based pooling algorithm has been applied for malignancy-oriented features in the CNN for breast cancer classification using mammography images [31]. The authors proposed two new pooling algorithms: region base group-max pooling (RGP) and global group max-pooling (GGP). Mathematically, RGP and GGP are represented by equations (5.1) and (5.2).

$$v = \frac{1}{K}\sum_{k=1}^{K} \tilde{z}_k \tag{5.1}$$

where $\tilde{z}_k$ represents the feature of the region $\tilde{s}_k$.

$$v_c = \frac{1}{K}\sum_{k=1}^{K} \tilde{z}_k^c \tag{5.2}$$

where v_c represents the cth component of final feature v. v_c is calculated as the mean values of the chosen portions of channel c.

The proposed method in their work was verified on the INbreast and CBIS datasets. The best results were obtained with various combinations of pooling algorithms and datasets, RGP + INbreast, RGP + CBIS, GGP + INbreast, and GGP + CBIS, providing accuracy of 0.923 ± 0.0003, 0.762 ± 0.0002, 0.922 ± 0.0002, and 0.767 ± 0.0002 respectively. The application of such a method can be verified on highly dense histopathology images for breast cancer detection and classification.

A U-net threshold map (TM) layer-based 3D CNN model was designed with threshold loss for automatic cancer detection from automated breast ultrasound (ABUS) images [32]. The threshold loss introduced in their work is given by equation (5.3).

$$L_{threshold}(W, w^N) = 1 - \frac{2 * \left| Mask(y_i = 1 \mid X; W, w^N) * Y \right|}{\left| Mask(y_i = 1 \mid X; W, w^N) \right| + |Y|} \tag{5.3}$$

where

$$Mask(y_i \mid X; W, w^N) = 1 / (1 + e^{-tmp})$$ (5.4)

and

$$tmp = \begin{cases} P(y_i = 1 \mid X; W, w^N), P(y_i) > TM(y_i) \\ -e^{10}, Otherwise \end{cases}$$ (5.5)

w^N = weight of *TM* layer, and W = weight of the main network.

The proposed model in their work provided a sensitivity of 95% and false-positive per volume of 0.84.

Mitotic cell count describes the development of cancerous tissue in the breast. Mitotic cell-based segmentation, detection, and classification methods have been adopted by authors [33]. The proposed model in their work has been termed MitosisNet by the authors. The segmentation, detection, and classification from whole slide images are performed by regression models based on R2U-Net and R2U-Net and IRRCNN models. The authors tested their method on MITOSIS-12, MITOSIS-14, and CWRU datasets. The best result was obtained by the combination of MitosisNet + multi patch on the MITOSIS-12 dataset, i.e., F1-score of 0.878.

Deep learning-based mask generation was adopted for the detection of epithelial cells from breast cancer stained with Ki-67, estrogen receptor (ER), and progresterone receptor (PR). The images were trained by a partially trained VGG-16 CNN model [34]. The area under the curve of mean receiver operating characteristic (ROC) was calculated to evaluate the model and the result was 0.93 for this designed model.

Hematoxylin and eosin-stained breast cancer and lymphoma histopathology images were processed using a deep learning-based autoencoder model for detection and classification [35]. The deep learning model was a residual autoencoder-based CNN model, a part of FusionNet. The projected model in their work was tested on the dataset available at [36] and provided 89.57% accuracy in invasive ductal carcinoma (IDC) detection and 97.67% for lymphoma classification.

Authors in [37] have proposed a cancerous mass detection and its classification as malignant and benign using CNN and extreme learning machine (ELM). After mass detection, a fusion of features was formed consisting of deep, texture, density, and morphological features. The classification was performed using the ELM model trained with the fused features. The ELM model is a feed-forward neural network developed by Huang *et al.* [38]. The method provided 86.50% classification accuracy.

The application of a fully connected CNN-based autoencoder has been carried out in [39] for breast histopathology image classification. One class SVM model was used to create anomalies in generated patches. Then classification was done using CNN. The proposed method in their work was verified on the biopsy image of breast cancer stained with hematoxylin and eosin and provided by the Israel Institute of Technology [40]. Still, the accuracy obtained by this method, i.e. 76%, needs to be increased to compete with models providing greater accuracy.

Breast tumors of malignant nature can move towards cancer development. Tumor detection is another part of breast cancer detection. For this purpose, a 3D CNN model has been utilized for processing and classification of breast ultrasound images [41]. First, a sliding windowing technique was adopted for extraction of the volume of interest (VOI) regions. These VOIs were then passed to the 3D CNN model for feature extraction and classification; 95% sensitivity was observed in this method.

A deep active learning-based mechanism has been adopted for labeling data and further classification of breast cancer histopathological images [42]. The model was designed utilizing two selection strategies: entropy-based strategy and confidence-boosting strategy. High entropy and high confidence were the main strategies to choose valuable samples. Then the updated training samples were used for training and classification using a pre-trained AlexNet model. The method discussed in their work was verified on the BreastHis dataset. Image- and patient-level accuracy of 91.61% and 92.84% were obtained.

Breast cancer classification has been performed from gene expression data in [43] using a lightweight CNN model. The gene expression data was used to form 2D images that were used for further training and classification. The reconstructed images were free from any type of complex shapes, therefore, the CNN was termed as lightweight. The proposed model was verified on RNA-Seq gene expression data from Pan-Cancer Atlas and provided 98.76% accuracy. The method involved the conversion of gene expression data to the image, which might be avoided using 1D CNNs to avoid such preprocessing.

Another work has been developed for breast cancer detection from mammograms and tomosynthesis images using the CNN model [44]. The base of the model was developed using the data augmentation technique and transfer learning algorithm. Data augmentation was designed by flipping the original image horizontally and then by rotating the image by 90°, 180°, and 270°. A transfer learning algorithm was implemented using the AlexNet, pre-trained with the ImageNet dataset. The model was evaluated using the measure area under ROC (auROC). The model provided auROC of 0.6749 and 0.7237 for mammograms and tomosynthesis classification.

The combination of deep learning and machine learning techniques is the new trend in the data processing. A combined model formed with pre-trained deep CNN (DCNN) models and ensemble SVM has been proposed for the classification of breast histopathology images into four classes, i.e., normal, invasive carcinoma, benign, and in situ carcinomas [45]. The authors used four deep learning-based pre-trained models, DenseNet, ResNet, VGG-16, and Inception V3 for feature generation. The generated features were then processed for feature selection using dual-network orthogonal low-rank learning (DOLL) so that these selected features can be utilized to train the SVM model, achieving 97.70% accuracy. Although deep learning methods were adopted in their work, the authors passed the ICACR-18 dataset through preprocessing steps like scale transformation and then color enhancement before passing to DCNN models.

Mitotic cell detection from breast cancer histopathology images has been carried out using two CNN models connected in parallel [46]. One of the two CNNs was trained with weak labels and the other model was trained with strong labels. The labels assigned to centroid pixels were known as weak labels and the labels assigned at pixel level were known as strong labels. System performance was checked on the datasets ICPR 2014 and AMIDA13, providing F-score values of 0.575 and 0.698 respectively. The prediction using such methods needs to be improved.

The combination of convolutional layers and recurrent layers is another option to design hybrid models and this is termed a CRNN model. A CRNN model has been used for breast ultrasound image classification to detect the presence of benign or malignant breast mass [47]. Before applying the CRNN model, authors processed the Mendeley dataset through extended line segment feature analysis (eLFA). Using this framework authors achieved 99.75% classification accuracy.

Breast cancer detection using a stacked ensemble of three CNN models was utilized in [48]. The authors detected the presence of cancer from 1D gene data and also from the breast histopathology image data. 1D gene data were first converted to images using DeepInsight [49] framework using mean normalization. Images of both datasets were passed through empirical wavelet transform and variational mode decomposition models for molecular feature extraction. The stacked ensemble model performed well in this work.

From the above study, it is observed that deep learning-based techniques outperform machine learning techniques in both single and hybrid categories. The datasets are of different forms depending upon the imaging techniques.

5.3 PROPOSED METHOD

Deep learning is an emerging technique in the field of image processing for its high computational performance. From the literature survey section, it can be observed that most of the works have CNN as the feature extractor as well as a classifier. Other forms of deep learning techniques such as recurrent neural network (RNN), long short-term memory (LSTM), and gated recurrent unit (GRU) also need to be explored in the field of breast cancer detection and classification. For this purpose, authors have considered the LSTM as the base model and formed a stacked ensemble model in combination with MLP for IDC, and non-IDC breast histopathology image classification. The workflow diagram of the proposed model is provided in Figure 5.2.

5.3.1 The Long Short-Term Memory

LSTM-based models are trained using a backpropagation algorithm. LSTMs are capable of handling long-term dependencies on the sequence of data [50]. An image is a row and column-wise arrangement of pixels. The CNN models are operated on images with the use of convolutional filters, but in LSTM, the pixels are first converted to time-dependent sequence data and then processed for feature extraction and classification. The structure of the LSTM cell is shown in Figure 5.3.

The output of the sigmoid activation function is related to the corresponding input as provided in equation (5.6):

$$f_t = \sigma(W_f[h_{t-1}, x_t] + b_f) \tag{5.6}$$

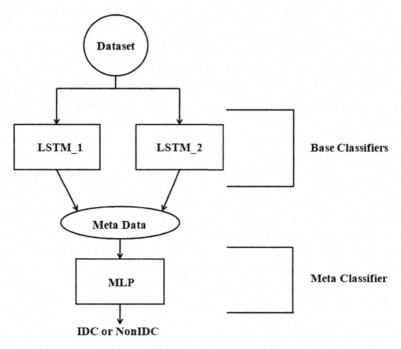

FIGURE 5.2   Block diagram of the proposed method.

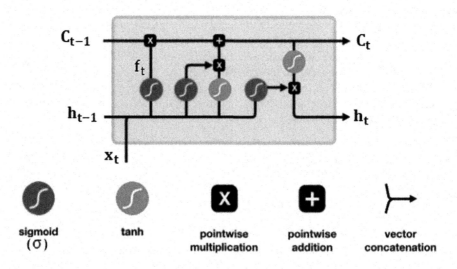

FIGURE 5.3   The internal structure of long short-term memory cell.

where W_f represents the weight vector of each cell, h_{t-1} is the result taken from the previous timestamp, x_t denotes the input at time t, and b_f represents the bias.

The output of the second sigmoid activation function and tanh activation function are represented in equations (5.7) and (5.8).

$$i_t = \sigma(W_i[h_{t-1}, x_t] + b_i) \tag{5.7}$$

$$C_t^{\sim} = \tanh(W_c[h_{t-1}x_t] + b_c, \ C_t = f_t * C_{t-1} + i_t * C_t. \tag{5.8}$$

where C_t, (W_i, W_c), and (b_i, b_c) represent the cell state, weights, and bias respectively.

The output of the last sigmoid activation function and tanh activation functions are provided in equations (5.9) and (5.10).

$$O_t = \sigma(W_o[h_{t-1}, x_t] + b_o) \tag{5.9}$$

$$h_t = O_t * \tanh(C_t) \tag{5.10}$$

where O_t represents the output, W_o is used for weight, and b_o represents the bias used in the output section.

In this work, the classification was performed by considering two different models. First, a single LSTM model was designed for training and classification. Then the result obtained from a single model was increased by training two LSTMs in parallel and passing the result of such an ensemble model to a stacked MLP model.

The ensemble of two LSTMs was developed using algorithm 1.

Algorithm 1: Training of base classifiers

1. Input = dataset
2. Input = nparray(input)
3. For $i = 1$ to 2
   Train $LSTM_i$ with input
4. $y = concatenate(\hat{y}_i)$
5. Save $LSTM_i$

where $\hat{y}_i$ represents the prediction of each model and y represents the concatenation of prediction of each model.

5.3.2 MULTI-LAYER PERCEPTRON

The concatenation of the predictions from base classifiers was transferred to the final classifier MLP, named as meta classifier. The MLP model was designed with three layers, namely input layer, hidden layer, and output layer. The input layer was of size 4 to accept the concatenation of two base classifiers with two output nodes. The hidden layer was formed with 32 nodes activated by the ReLU activation function and the output layer consisted of two nodes activated by the Softmax activation function

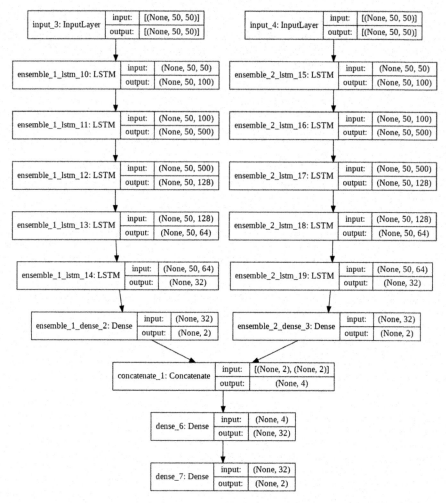

FIGURE 5.4 The proposed model with internal parameters.

for final classification. The system parameters of the proposed model are shown in Figure 5.4.

5.4 RESULTS AND DISCUSSION

5.4.1 DATASET

The proposed model was verified using the breast histopathology images distinguishing non-IDC and IDC images of size 50 × 50 pixels. The dataset is publicly available at the online platform Kaggle [51]. The original breast histopathology dataset consisted of 162 numbers of whole-mount slide images scanned at 40×. The IDC and non-IDC patches of size 50×50 were extracted with labels containing the patient ID, location of pixels in x and y direction from which the patches were cropped, and the

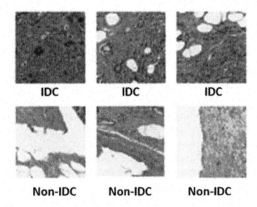

FIGURE 5.5   Samples of training data.

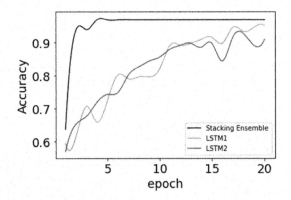

FIGURE 5.6   Validation accuracy comparison.

corresponding class. Class value 0 represented non-IDC and 1 represented IDC. After collecting the dataset we have renamed the dataset images with simple labels, such as IDC for images containing cancer cells and non-IDC for healthy breast histopathology images. The samples of each category of breast histopathology images are shown in Figure 5.5.

5.4.2   PREDICTION RESULTS

Prediction by the single LSTM was found to be less in comparison to state-of-the-art methods. The accuracy plots of the two single models along with the ensemble model are shown in Figure 5.6. To increase performance, the ensemble learning algorithm stacked with MLP was applied. Accuracy increased, reaching 98%.

The proposed method was evaluated using the parameters precision, recall, F1-score, accuracy, macro average, and weighted average. The detailed values of these parameters are provided in Table 5.1.

TABLE 5.1
Performance parameters of the proposed model

	Precision	Recall	F1-score	Accuracy	Support
Non-IDC (0)	1.00	0.93	0.96	0.98	14
IDC (1)	0.97	1.00	0.98		28
Macro average	0.98	0.96	0.97		42
Weighted average	0.98	0.98	0.98		42

Non-IDC, non-invasive ductal carcinoma; IDC, invasive ductal carcinoma.

From Figure 5.6, it is observed that the accuracy due to stacked ensemble increased in comparison to the single LSTM model. The increase in performance is due to second-stage training with the predictions of first-stage outputs.

5.5 CONCLUSION

In this chapter, the authors have provided a short review of breast cancer detection and classification using machine learning and currently trending deep learning techniques. The study provides a brief idea of the classification of different types of datasets, including mammograms, MRI data, CT scans, and ultrasound, thermography, histopathology, and microwave images. In recent years, machine learning techniques have also been adopted, although in an ensemble manner. Also, deep learning in single architecture is rarely used. Ensemble learning has proven to give a better performance and the methods were discussed in this chapter. Also, the authors have verified an LSTM-based model for the classification of breast histopathology images to predict non-IDC versus IDC images. Research in medical applications is an ongoing process. More advanced algorithms are being developed for better classification. Still, the scope is there to design models with better performance, practicability, and cost-efficiency.

REFERENCES

[1] *IAFR Cancer. Global Cancer Observatory.* 2020. Available from: http://gco.iarc.fr/.
[2] *Division of Cancer Prevention and Control, Centers for Disease Control and Prevention.* Available from: www.cdc.gov/cancer/breast/basic_info/symptoms.htm.
[3] Kennedy, D.A., T. Lee, and D. Seely, A comparative review of thermography as a breast cancer screening technique. *Integrative Cancer Therapies* 2009; **8**(1): 9–16.
[4] Hossam, A., et al., Performance analysis of breast cancer imaging techniques. *International Journal of Computer Science and Information Security (IJCSIS)* 2017; **15**(5).
[5] Bhidé, A., et al., Case histories of significant medical advances: Gastrointestinal endoscopy. *Harvard Business School Accounting and Management Unit Working Paper* 2019 (20-005).
[6] Na, S.P. and D. Houserkovaa, The role of various modalities in breast imaging. *Biomed Pap Med Fac Univ Palacky Olomouc Czech Repub* 2007; **151**(2): 209–218.

[7] Kurmi, Y., et al., Tumor malignancy detection using histopathology imaging. *Journal of Medical Imaging and Radiation Sciences* 2019; **50**(4): 514–528.

[8] Amalu, W.C., *A Review of Breast Rhermography.* 2002: pp. 1–21. Available from: https://clinicalthermography.co.nz/A_review_of_Breast_Thermography.php

[9] Zhang, H.J.I.A., Microwave imaging for breast cancer detection: The discrimination of breast lesion morphology. *IEEE Access* 2020; **8**: 107103–107111.

[10] Alibakhshikenari, M., et al., Metamaterial-inspired antenna array for application in microwave breast imaging systems for tumor detection. *IEEE Access* 2020; **8**: 174667–174678.

[11] Mehranpour, M., et al., Robust breast cancer imaging based on a hybrid artifact suppression method for early-stage tumor detection. *IEEE Access* 2020; **8**: 206790–206805.

[12] Sharma, S., A. Aggarwal, and T. Choudhury, Breast cancer detection using machine learning algorithms. In *2018 International Conference on Computational Techniques, Electronics and Mechanical Systems (CTEMS)*. 2018. IEEE.

[13] Tahmooresi, M., et al., Early detection of breast cancer using machine learning techniques. *Journal of Telecommunication, Electronic and Computer Engineering (JTEC)* 2018; **10**(3–2): 21–27.

[14] Alarabeyyat, A. and M. Alhanahnah, Breast cancer detection using k-nearest neighbor machine learning algorithm. In *2016 9th International Conference on Developments in eSystems Engineering (DeSE)*. 2016. IEEE.

[15] Gayathri, B., et al., Breast cancer diagnosis using machine learning algorithms-a survey. *International Journal of Distributed and Parallel Systems* 2013; 4(3): 105.

[16] Mambou, S.J., et al., Breast cancer detection using infrared thermal imaging and a deep learning model. *Journal Sensors* 2018; **18**(9): 2799.

[17] Shen, L., et al., Deep learning to improve breast cancer detection on screening mammography. *Scientific Reports* 2019; **9**(1): 1–12.

[18] Ragab, D.A., et al., Breast cancer detection using deep convolutional neural networks and support vector machines. *PeerJ* 2019; **7**: e6201.

[19] Wang, D., et al., Deep learning for identifying metastatic breast cancer. *arXiv preprint* 2016.

[20] Chiu, H.-J., T.-H.S. Li, and P.-H. Kuo, Breast cancer–detection system using PCA, multilayer perceptron, transfer learning, and support vector machine. *IEEE Access* 2020. 8: p. 204309–204324.

[21] Rashmi, R., et al., A comparative evaluation of texture features for semantic segmentation of breast histopathological images. *IEEE Access* 2020. **8**: p. 64331–64346.

[22] Li, G., et al., Effective breast cancer recognition based on fine-grained feature selection. *IEEE Access* 2020; **8**: 227538–227555.

[23] Huang, Q., et al., On combining biclustering mining and AdaBoost for breast tumor classification. *IEEE Transactions on Knowledge and Data Engineering* 2019; 32(4): 728–738.

[24] Acharya, U.R., et al., A novel algorithm for breast lesion detection using textons and local configuration pattern features with ultrasound imagery. *IEEE Access* 2019; **7**: 22829–22842.

[25] Beevi, K.S., et al., A multi-classifier system for automatic mitosis detection in breast histopathology images using deep belief networks. *IEEE Journal of Translational Engineering in Health and Medicine* 2017; **5**: 1–11.

[26] Chen, R., et al., A stacked autoencoder neural network algorithm for breast cancer diagnosis with magnetic detection electrical impedance tomography. *IEEE Access* 2019; **8**: 5428–5437.

[27] Elston, E.W. and I.O. Ellis, Method for grading breast cancer. *Journal of Clinical Pathology* 1993; **46**(2): 189.
[28] Kausar, T., et al., Small mitosis: Small size mitotic cells detection in breast histopathology images. *IEEE Access* 2020.
[29] Das, K., et al., Detection of breast cancer from whole slide histopathological images using deep multiple instance CNN. *IEEE Access* 2020; **8**: 213502–213511.
[30] Man, R., P. Yang, and B. Xu, Classification of breast cancer histopathological images using discriminative patches screened by generative adversarial networks. *IEEE Access* 2020; **8**: 155362–155377.
[31] Shu, X., et al., Deep neural networks with region-based pooling structures for mammographic image classification. *IEEE Transactions on Medical Imaging* 2020; **39**(6): 2246–2255.
[32] Wang, Y., et al., Deeply-supervised networks with threshold loss for cancer detection in automated breast ultrasound. *IEEE Transactions on Medical Imaging* 2019; **39**(4): 866–876.
[33] Alom, M.Z., et al., MitosisNet: End-to-end mitotic cell detection by multi-task learning. *IEEE Access* 2020; **8**: 68695–68710.
[34] Valkonen, M., et al., Cytokeratin-supervised deep learning for automatic recognition of epithelial cells in breast cancers stained for ER, PR, and Ki-67. *IEEE Transactions on Medical Imaging* 2019; **39**(2): 534–542.
[35] Brancati, N., et al., A deep learning approach for breast invasive ductal carcinoma detection and lymphoma multi-classification in histological images. *IEEE Access* 2019; **7**: 44709–44720.
[36] *Lymphoma and IDC Datasets. Accessed: 2016.* Available from: www.andrew janowczyk.com/deep-learning/
[37] Wang, Z., et al., Breast cancer detection using extreme learning machine based on feature fusion with CNN deep features. *IEEE Access* 2019; **7**: 105146–105158.
[38] Huang, G.-B., Q.-Y. Zhu, and C.-K. Siew, Extreme learning machine: theory and applications. *Neurocomputing* 2006; **70**(1–3): 489–501.
[39] Li, X., et al., Discriminative pattern mining for breast cancer histopathology image classification via fully convolutional autoencoder. *IEEE Access* 2019; **7**: 36433–36445.
[40] *Breast Cancer Imageset. [Online].* Available from: ftp://ftp.cs.technion.ac.il/pub/projects/medicimage/breastcancerdata/
[41] Chiang, T.-C., et al., Tumor detection in automated breast ultrasound using 3-D CNN and prioritized candidate aggregation. *IEEE Transactions on Medical Imaging* 2018; **38**(1): 240–249.
[42] Qi, Q., et al., Label-efficient breast cancer histopathological image classification. *IEEE Journal of Biomedical and Health Informatics* 2018; **23**(5): 2108–2116.
[43] Elbashir, M.K., et al., Lightweight convolutional neural network for breast cancer classification using RNA-seq gene expression data. *IEEE Access* 2019; **7**: 185338–185348.
[44] Zhang, X., et al., Classification of whole mammogram and tomosynthesis images using deep convolutional neural networks. *IEEE Transactions on Nanobioscience* 2018; **17**(3): 237–242.
[45] Wang, Y., et al., Breast cancer image classification via multi-network features and dual-network orthogonal low-rank learning. *IEEE Access* 2020; **8**: 27779–27792.
[46] Sebai, M., T. Wang, and S.A. Al-Fadhli, PartMitosis: A partially supervised deep learning framework for mitosis detection in breast cancer histopathology images. *IEEE Access* 2020; **8**: 45133–45147.
[47] Kim, C.-M., R.C. Park, and E.J. Hong, Breast mass classification using eLFA algorithm based on CRNN deep learning model. *IEEE Access* 2020; **8**: 197312–197323.

[48] Das, A., M.N. Mohanty, P.K. Mallick, P. Tiwari, K. Muhamad, and H. Zhu. Breast cancer detection using an ensemble deep learning method. *Biomedical Signal Processing and Control* 2021; **70**: 103009.

[49] Sharma, A., E. Vans, D. Shigemizu, K.A. Boroevich, and T. Tsunoda. (2019). DeepInsight: A methodology to transform a non-image data to an image for convolution neural network architecture. *Scientific Reports* **9**(1): 1–7.

[50] Das, A., G.R. Patra, and M.N. Mohanty. (2020, July). LSTM based Odia handwritten numeral recognition. In *2020 International Conference on Communication and Signal Processing (ICCSP)* (pp. 0538–0541). 2020. IEEE.

[51] Janowczyk, A., and A. Madabhushi. (2016). Deep learning for digital pathology image analysis: A comprehensive tutorial with selected use cases. *Journal of Pathology Informatics* **7**: 29.

6 Artificial Intelligence and Machine Learning
A Smart Science Approach for Cancer Control

Kaushik Dehingia
Gauhati University, Guwahati, Assam, India

Mdi B. Jeelani
Imam Muhammad ibn Saud Islamic University, Riyadh, Saudi Arabia

Anusmita Das
Gauhati University, Guwahati, Assam, India

6.1 INTRODUCTION

Artificial intelligence (AI) is a branch of computer science that was developed in the early twentieth century to solve complex problems using computing systems with advanced analytical or predictive capabilities. One of the most promising branches of AI is machine learning (ML), which can learn from past data and has the potential to create programming languages from that data. ML techniques forecast problem-specific outcomes and measurements based on learning ability [Kourou et al. 2015, Jiang et al. 2020]. ML can "learn" from data and recognize sophisticated patterns using a mathematical and statistical approach. When computing a complex dataset, ML processes go through two stages. The first is referred to as the "learning" phase, while the second is the "verification" phase. During the first phase's implementation, a theoretical model is developed to clarify the task. ML algorithms check the first phase's findings and details during the verification phase. ML algorithms are divided into three groups during the learning phase: supervised, unsupervised, and reinforcement learning [Thenault et al. 2020].

Deep learning (DL) is a sub-branch of ML, in which artificial neural networks (ANNs), convolutional neural networks (CNNs), and deep neural networks (DNNs) are used to analyze the deep data layers to automate the removal and recognition of patterns in big datasets. The DL method is programmed to perform several tasks, such as computer vision, speech recognition, bioinformatics, drug design, medical image analysis, and histopathological diagnosis, with better performance than human intelligence [Coccia 2019, Leatherdale and Lee 2019].

DOI: 10.1201/9781003230540-6

ANNs consist of supervised and unsupervised ML algorithms programmed to mimic biological neurons' behavior. They perform the same functions as a biological neural network. In ANNs, there are several layers. The first layer is made up of nodes that link to form a processing network. There is another layer called hidden nodes between the first and last layers. The multilayer perceptron has three neuronal layers, and this is the most commonly used ANN in medical science. Magnetic resonance imaging (MRI), mammography, ultrasound, and thermography are the most common applications [Thenault et al. 2020].

In the twenty-first century, the mortality rate of cancer increased due to improper cancer management and diagnosis. Cancer is a group of diseases characterized by the irregular growth of normal tissues that spread to other body areas. Early cancer prediction and identification continue to pique the interest of AI and ML researchers, oncologists, and scientists due to the unpredictable growth of tumor cells. The early detection and diagnosis of cancer are critical in the battle against cancer. Image processing, which is used for cancer screening and diagnosis, has used various ML methods [Sherbet et al. 2018]. We will explain how different types of ML techniques and algorithms are used to detect cancer through medical imaging, monitor cancer care such as radiotherapy, and reduce the cancer burden in this study.

6.2   TYPES OF DATA

The vital aim of the ML method is to identify and give rise to a model which can perform categorization prognosis approximation and related queries. Hence for an excellent model to work, we must test its accuracy. For this purpose, quantitative metrics of accuracy and area under the curve (AUC) are used. Thus, AUC is considered the mode to assess the model's performance; it is measured in terms of sensitivity and specificity (Figure 6.1).

Predictive accuracy depends on the data provided for training and testing the samples, hence the need for a classifier. The classification of data is based on these methods:

- Method of holdout
- Sampling the random data
- Method of cross-validation
- Method of bootstrap.

The samples were categorized as training sets and test sets during the holdout procedure. Hence the categorization model opted for training sets, and its execution was approximated on the test sets.

In the random method, we performed the same categorization as in the holdout method, where the training sets and test sets were selected randomly, and the holdout method was repeated several times.

Each specimen was utilized for training and only once for test sets in the cross-validation approach.

In the bootstrap method, the specimens were abstracted by replacing them with training and test sets.

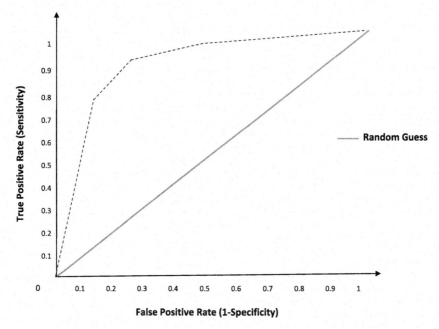

FIGURE 6.1 Area under the curve for measured performance of the model [Kourou et al. 2015].

6.2.1 ML METHODS

After processing the data, the new learning task must be given in the ML method. The ML methods, which we will discuss in the next section, are:

- ANN
- SVM.

6.2.2 ANN METHOD

This network has several distinct layers, consisting of an input layer, multiple hidden layers, and an output layer (Figure 6.2).

Consider the example of the vector transfer given in https://towardsdatascience. com/applied-deep-learning-part-1-artificial-neural-networks-d7834f67a4f6, which performs dot products of the matrices (Figure 6.3).

Here the vector x has 1 row and 3 columns and the output layer has 1 row and 1 column. Using the ANN technique, we use backpropagation, as shown in Figure 6.4. Here we perform matrix multiplication between 3 branches and 4 branches of the 1×3 matrix input with 3×4 weight matrix W_1, which is then multiplied by W_2 (4×4) arising in the 1×4 matrix. Lastly, we use the 4×1 matrix into W_3, to get output 4.

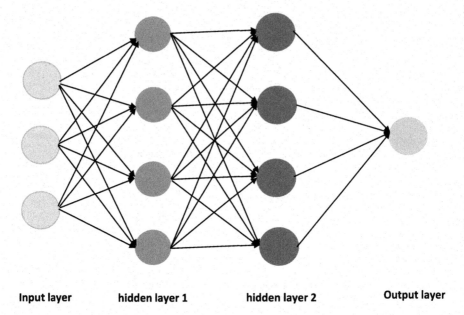

Input layer   **hidden layer 1**   **hidden layer 2**   **Output layer**

FIGURE 6.2   Network in which each branch in one layer is attached to another branch in the next layer [Kourou et al. 2015].

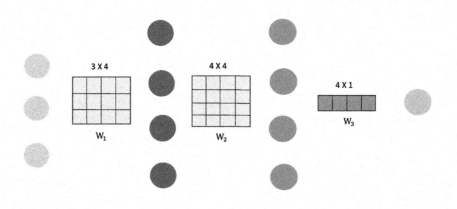

Input layer   **hidden layer 1**   **hidden layer 2**   **Output layer**

FIGURE 6.3   Dot product of the matrices in the form of layers in the artificial neural network model [Kourou et al. 2015].

6.2.2.1   Advantages

- Representation is made in pairs.
- They are used for problems of target function.
- The errors do not affect the final output.

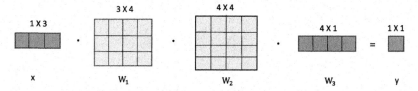

FIGURE 6.4 Back propagation for matrix multiplication [Kourou et al. 2015].

- It is used for quick evaluation. These methods can work with a number of parameters.

6.2.2.2 Disadvantages
- ANN requires parallel processing power for its structure, which makes it dependent on the equipment.
- When this method gives us a solution, it can give unexplained functionality of the network.
- It does not give optimum results.

6.2.2.3 ANN Methods in Cancer Control
One of the essential tools in medical diagnosis and tumor classification is to record gene appearance levels from several genes. For this purpose, an ANN method plays a significant role as we know that these data are indicated by a large number of gene appearance levels that can overlap with the number of specimens, giving rise to outfitting the data. Thus arises the need to use an appropriate method or select a small set of genes or implement cross-validation to avoid outfitting the data.

In Tibshirani et al. [2002], both supervised and unsupervised methods were used to classify cancer. The authors took taken an example of colon cancer where they proved that supervised ANNs were better than other methods. Colon cancers were classified into two subtypes: inflammatory bowel disease-related neoplasia (IBDN) and sporadic colon adenoma cancers (SACs). SACs require polypectomy, and IBDNs are treated with partial colectomy. A study was performed with DNA microchips to separate the usage of ANN and hierarchically combined colon cancer approaches using 8,064 hybridizing clones of cDNA in 39 neoplastic colon samples [Selaru et al. 2002]. The method of Gene Finder was considered for the determination of 1,192 clones that indicate various mean statistical square values between IBDN and SAC ($P = 0.001$). The authors investigated 1,192 inputs for the selected genes, and results were compiled at 0 for IBDNs and 1 for SACs with MATLAB mathematical software. The ANN method was accomplished with a training set of 5 IBDNs and 22 SACs. The test set had the remaining data samples, consisting of 3 IBDNs and 9 SACs. ANN methods employed statistical techniques, comparing their predefined output (target) with ANN output. Hierarchical congregated methods used the program Cluster. Here the ANN network offered 12 out of 12 blinded samples, and the researchers observed that the hierarchical congregated analysis deteriorated. Thus, ANN correctly recognized the two forms of colon cancer [Selaru et al. 2002].

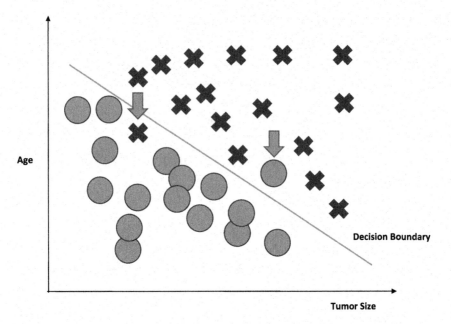

FIGURE 6.5 Tumors were differentiated according to their size and the age of the patient. Arrows represent misclassified tumors. This figure is modified from the machine learning lectures of Adams [2012].

6.2.3 SVM Method

In Huang et al. [2018], SVM was considered to be the most powerful tool/method for categorizing cancer cells. This method helps us understand the genomic patterns of different types of cancer, prognosis, prognostication for drugs, and the malignancy phenomena of cancer on tumor cells. In this method, a partition is created, known as a hyper-plane or decision boundary between two classes (Figure 6.5).

6.2.3.1 SVM Method in Cancer Control

To utilize this method, we took the example of using the SVM method in the early prognosis of breast cancer. The prognosis of tumors in mammograms is divided into three main stages. In the first stage, an enhancement method is carried out: image enhancement techniques are used to modify and clear an image. In the next stage, an intensity adjustment is made to a new range. After this, mammogram enhancement classifies the tumor area. Then these are taken from the segmented mammogram. Finally, in the next stage comes these segments are analyzed by the SVM classifier.

In this method, we considered the nearest points, which are known as support vectors.

For the given set of points $(a_1, b_1), (a_2, b_2)\ldots\ldots\ldots\ldots (a_n, b_n)$

a_i in **R** and b_i in $(-1, +1)$
a_i gives us vector notation and b_i gives us categorized sections.

Now the hyper-plane is defined as:

$$V_a^T + b = 0$$

where V = weighed vector

$$Va_i^T + C \geq +1 \qquad \text{if } b_i = +1$$

$$Va_i^T + C \leq -1 \qquad \text{if } b_i = -1$$

where C is the bias

Hence the objective function of the SVM method maximizes the $\dfrac{1}{\|V\|^2}$

where a_i for which $|b_i| Va_i^T + b = -1$ is named as a support vector.
The application of this method for breast cancer detection is explained in detail by Hiremath and Prasannakumar [2017].

6.2.3.1.1 Advantages
- This method is effective in high-dimensional space.
- It has a high efficiency rate in terms of memory.

6.2.3.1.2 Disadvantages
- It is not appropriate for large datasets.
- When numbers of data points are more than the sample data it does not give appropriate results.

In the following sections, we will gain insight into the application of ML methods in different types of cancer.

6.3 ROLE OF ML IN DIFFERENT CANCERS

Pathologists have always struggled with cancer identification, as well as diagnosis and treatment preparation. AI and ML models can detect cancer from microscopic biopsy images using automated detection and classification methods. The software loads microscopic biopsy images from a file. A CNN algorithm can read and arrange images through image processing. Images are trained and tested using ML, which examines the images and determines whether they are positive or negative. This approach is faster and produces even better results.

Cancer is a term used to identify cancerous growths or irregular cell growth in the body. There are over 100 different forms of cancer. Breast cancer is a form of cancer that affects the tissue of the breast. Breast cancers are most commonly found in the cells that line the milk ducts and the lobules that supply milk to these vessels. A biopsy of the connective tissue is used to diagnose breast cancer [Breast Cancer Treatment 2014]. According to the Canadian Cancer Society, 27,700 women and 260 men will be diagnosed with breast cancer in 2021 [Breast cancer statistics – Canadian Cancer Society].

A brain tumor is an intracranial tumor, an irregular mass of tissue in the brain that grows uncontrollably [Brain Tumors 2021]. The initial stage of tumors that start in the brain and secondary tumors that spread from tumors outside the brain are the two types of cancerous tumors. Symptoms of all types of brain tumors may arise, and they differ based on the location of the tumor in the brain [NCI 2014].

The unregulated proliferation of cancer cells in the skin is known as skin cancer. Basal cell carcinoma (BCC), squamous cell carcinoma (SCC), and melanoma are the three forms of skin cancer. Non-melanoma skin cancers are BCC and SCC. BCC is ubiquitous in the head and neck, where it develops slowly and affects the tissues around it. SCC affects cells in the outer layer of the epidermis.

Lung cancer is a malignant tumor of the lungs that causes uncontrolled cell growth. This growth spreads to surrounding tissues through a process called metastasis. Small-cell lung carcinoma and small-cell non-lung carcinoma are the two most common forms of lung cancer [Falk and Williams 2010].

Prostate cancer affects the prostate gland; the prostate gland is present in the male reproductive system and surrounds and protects the bladder. The majority of prostate cancers develop slowly. Cancer cells can spread to other body parts, especially the bones and lymph nodes [Ruddon 2007]. In 2021, an estimated 248,530 men will be diagnosed with this type of cancer, according to research from the American Cancer Society (www.cancer.org/cancer/prostate-cancer/about/key-statistics.html).

6.3.1 ROLE IN BRAIN TUMOR

When a brain tumor develops, most abnormal cells have poor contrast compared to neighboring cells. The majority of patients with brain tumors have a 5-year survival record. As a result, the most important job is to identify a brain tumor as soon as possible. Optical imaging and AI help to speed up and improve the identification and diagnosis of brain tumors. In medical imaging, AI in the application of DL methods is used to categorize and diagnose brain tumors. The most widely used MRI modality for brain tumor diagnosis is a painless technique that helps study tumors from different perspectives and viewpoints. The medical imaging method of brain tumor segmentation is used for tissue quantification, classification, surgical preparation, abnormality localization, and various medical assessments. Manual segmentation, semiautomatic segmentation, and fully automatic segmentation are the three forms of segmentation. Different AI and ML algorithms are used in MRI image processing to better view image segmentation and classification [Saba 2020]. In the segmentation of brain tumors, the CNN algorithm is widely used. With limited pre-processing, CNN extracts features from pixel images directly [Badža and Barjaktarovi 2020]. The best-known CNN architectures for brain tumor segmentation are given as fuzzy clustering, the form of k-clustering, and Otsu and threshold approaches, as well as U-Net architecture [Bhandarim et al. 2020]. LinkNet is a semantic segmentation-focused light DNN architecture. This network is faster and more accurate than SegNet [Sobhaninia et al. 2018, Saba 2020].

6.3.2 ROLE IN BREAST CANCER

Breast cancer is common in women worldwide due to delayed childbearing, insufficient breast therapy, and a higher screening rate. *Mammography* is a low-cost and safe breast cancer screening technique that helps to enhance prognosis and minimize mortality by detecting unchecked breast cells early. Computer-aided detection (CADe) and diagnosis (CADx) imaging methods are used in this screening process. These imaging techniques use AI-based CNN models to provide accurate mammography information. To improve the accuracy of digital mammography, an AI-based algorithm is used. CNNs based on AI are often used to scan and examine microscopic images of cancer cells. Abnormal cells are isolated from the healthy cells during this procedure [Coccia 2019, Patil et al. 2020]. Deep synergy models are AI-based programs that predict novel drug combinations within the space of drugs and cell lines investigated [Akbari et al. 2019].

In recent years, DL algorithms have also been used to determine breast volume and parenchyma arrangement and essential biomarkers for cancer risk assessment and treatment management [Saba 2020]. The success of the classification system has had a significant impact on breast cancer diagnosis. The spread of breast cancer has been identified and listed using CNN algorithms AlexNet, CiFarNet, GoogLeNet, VGG16, and VGG19. Other ML techniques such as neural networks, decision trees, K-nearest neighbor (KNN), support vector machine (SVM), and ensemble classifiers are also used to differentiate between malignant and benign tumors [Kourou et al. 2015, Patil et al. 2020].

6.3.3 ROLE IN SKIN CANCER

The two most common types of skin cancer are melanoma and non-melanoma. Because of melanoma's high variability and propensity to spread, identification can be more difficult. Dermatologists can diagnose and classify skin cancer using visual examination and dermoscopy. They were able to identify the cancer stage and the treatment or diagnostic protocol for cancer using their pattern recognition experience. AI and ML algorithms are gaining much traction in skin cancer diagnosis [Goyal et al. 2020]. According to Esteva et al. [2017], the performance of DL algorithms in skin cancer classification is extremely strong compared to dermatologists' performance in skin cancer classification. ML technologies that extract sophisticated features, such as the ABCD rule, the three-point checklist, and deep CNN (DCNN), have also obtained substantial achievements in the area of medical imaging, in which features are directly derived from the images. Ramya et al. [2015] used an adaptive histogram equalization method, a Wiener filter, a segmentation mechanism, and an SVM classifier to identify a small dataset of skin lesions. The skin cancer categorization was highly reliable and widely accepted. Aima and Sharma [2019] achieved an accuracy of 74.76% and a validation loss of 57.56% in their work when handling early-stage melanogenic skin identification by CNN on 514 dermoscopic pictures from International Skin Imaging Collaboration (ISIC) datasets.

6.3.4 ROLE IN PROSTATE CANCER

Because of poor management during the diagnosis process, prostate cancer is the second leading cause of cancer. The detection of an elevated prostate-specific antigen (PSA) level during the diagnosis of prostate cancer is important. The PSA level was measured using an ANN-based algorithm, which was also used to diagnose prostate cancer and its progression. ANN has been used in several studies to increase the sensitivity and precision of prostate cancer detection, reduce unnecessary prognostic testing of false-positive PSA outcomes, and help in prostate cancer diagnosis. DL, an ML-based supervised methodology, is more effective for different imaging modalities and treatment and produces high-performance results in prostate cancer [Patil et al. 2020, Thenault et al. 2020]. Multi-parametric MRI (mp MRI) based on CNN is used to identify suspicious prostate lesions and gain insight into tissue properties. CNN algorithms combined with mp MRI images effectively detect and treat prostate cancer [Wildeboer et al. 2020]. Through advanced and versatile CADe and CADx systems, AI models recognize and classify prostate cancer.

6.3.5   ROLE IN LUNG CANCER

AI and ML-based approaches can also assist with lung cancer identification, diagnosis, and prognosis. Researchers used an AI algorithm in lung cancer classification to achieve better results in image analysis, especially in computed tomography (CT) scans. In the medical image analysis method, CNN-based models can learn directly from raw datasets and extract features automatically [Arabi and Zaidi 2020]. The multi-CNN (MCNN) model can capture nodular heterogeneity by removing discriminative features from alternately stacked layers in early lung cancer diagnosis. By analyzing CT images, this model can measure lung nodules as well. DL technology, such as CNN, assists in lung cancer diagnosis by achieving high accuracy in classifying the type of cancer and assessing the best course of treatment [Guosheng et al. 2020].

6.3.6   ROLE IN RADIOTHERAPY

Based on structured and unstructured radiation oncology datasets, ML applications for radiotherapy have been developed. The primary aim of modern radiotherapy is to synthesize helpful information from these data to enhance patient outcomes. Knowledge-based response-optimized radiotherapy has since emerged as a critical framework for designing individualized therapies in the radiotherapy treatment course by modifying dose delivery based on clinical, geometric, and physical parameters. The application of AI technology during cancer radiotherapy increases. Radiologists may use AI to create radiation laws that automatically get them out of target areas or into treatment areas. The AI program creates a care plan for patients equivalent to conventional treatment plans, in a lot less time [Arabi and Zaidi 2020, Guosheng et al. 2020].

6.3.7   ROLE IN CHEMOTHERAPY

In chemotherapy drugs are administered to the body that destroy cancer cells while also halting the growth of abnormal cells. Chemotherapy uses potent chemicals, so it poses more of a risk than a profit. These adverse effects can be reduced to some degree with the aid of AI. AI can assist in identifying and analyzing cancer-causing genes, as well as identifying mutation level, cancer stage, and subtype based on mutation level [Arabi and Zaidi 2020]. Furthermore, an AI-based model can monitor chemotherapy drug use, determine drug tolerance, and rapidly understand how cancer cells become immune to anticancer drugs, thus improving drug production and aiding in drug regulation. The method of DL helps to correct and speed up optimizing combination chemotherapy [Artificial Intelligence Benefits in Chemotherapy – AI Objectives].

6.4 CONCLUSION

AI and ML are gaining popularity in biomedical research. They are commonly used in medical and health science research due to their well-programmed methods, effective system for data processing and optimization, and insightful, accurate performance. Cancer is one of the most common diseases in human society, killing millions of people worldwide. AI and ML algorithms are widely used in cancer research and treatment. In this work, we have discussed the applications of two key algorithms, ANN and SVM method, in cancer detection. Also, various data types used in ANN and SVM systems, such as the holdout method, random data sampling, cross-validation method, and bootstrap method, were discussed. The role of ML techniques such as DL methods, CNN, MCNN, and DCNN in treating cancers such as brain tumors, and breast, skin, prostate, and lung cancer was then addressed. These machine-learning methods can also be applied to the delivery of radiotherapy and chemotherapy, two of the most common cancer treatments.

REFERENCES

Adams, S. 2012. Is Coursera the beginning of the end for traditional higher education? Forbes. www.forbes.com/sites/susanadams/2012/07/17/is-coursera-the-beginning-of-the-end-for-traditional-higher-education/?sh=599ac54212ae.

Aima, A., Sharma, A.K. 2019. Predictive approach for melanoma skin cancer detection using CNN (March 14, 2019). Proceedings of International Conference on Sustainable Computing in Science, Technology and Management (SUSCOM), Amity University Rajasthan, Jaipur, India, February 26–28, 2019, Available at: https://ssrn.com/abstract=3352407

Akbari, P., Gavidel, P., Gardaneh, M. 2019. The revolutionizing impact of artificial intelligence on breast cancer management. *Archives of Breast Cancer*, 6(1): 1–3. DOI: 10.32768/abc.201961-3.

Arabi, H., Zaidi, H. 2020. Applications of artificial intelligence and deep learning in molecular imaging and radiotherapy. *European Journal of Hybrid Imaging*, 4: 17. https://doi.org/10.1186/s41824-020-00086-8

Badža, M.M., Barjaktarovi, M. C. 2020. Classification of brain tumors from MRI images using a convolutional neural network. *Applied Sciences*, 10: 1999. doi:10.3390/app10061999

Bhandarim, A., Koppen, J., Agzarian, M. 2020. Convolutional neural networks for brain tumour segmentation, *Insights into Imaging*, 11: 77. https://doi.org/10.1186/s13244-020-00869-4.

Brain Tumors – Classifications, Symptoms, Diagnosis and Treatments. www.aans.org. Retrieved 29 January 2021.

Breast cancer statistics – Canadian Cancer Society. Available from: https://cancer.ca/en/cancer-information/cancer-types/breast/statistics

Breast Cancer Treatment, NCI. 23 May 2014. Available from: www.ncbi.nlm.nih.gov/books/NBK65969/

Coccia, M. 2019. Artificial intelligence technology in oncology: A new technological paradigm. arXiv:1905.06871

Esteva, A., Kuprel, B., Novoa, R.A., Ko, J., Swetter, S.M., Blau, H.M., Thrun, S. 2017. Dermatologist-level classification of skin cancer with deep neural networks. *Nature*, 542 (7639): 115–118. https://doi.org/10.1038/nature21056

Falk, S., Williams C. 2010. *Chapter 1. Lung Cancer – The Facts* (3rd ed.): Oxford: Oxford University Press, pp. 3–4.

Goyal, M., Knackstedt, T., Yan, S., Hassanpour, S. 2020. Artificial intelligence-based image classification methods for diagnosis of skin cancer: Challenges and opportunities. *Computers in Biology and Medicine*, 127: 104065. https://doi.org/10.1016/j.compbiomed.2020.104065

Guosheng, L., Wenguo, F., Hui, L., Xiao, Z. 2020. The emerging roles of artificial intelligence in cancer drug development and precision therapy. *Biomedicine and Pharmacotherapy*, 128: 110255. doi.org/10.1016/j.biopha.2020.110255

Hiremath, B., Prasannakumar, S.C. 2017. Automated evaluation of breast cancer detection using SVM classifier. *International Journal of Computer Science Engineering and Information Technology Research (IJCSEITR)*, 5 (Issue 1): 7–16.

Huang, S., Cai, N., Pacheco, P.P., Narrandes, S., Wang, Y., Xu, W. 2018. Applications of support vector machine (SVM) learning in cancer genomics. *Cancer Genomics Proteomics*, 15(1): 41–51. doi: 10.21873/cgp.20063. PMID: 29275361; PMCID: PMC5822181.

Jiang, Y., Yang, M., Wang, S., Li, X., Sun, Y. 2020. Emerging role of deep learning-based artificial intelligence in tumor pathology. *Cancer Communications*, 40: 154–166. DOI: 10.1002/cac2.12012

Kourou, K., Exarchos, T.P., Exarchos, K.P., Karamouzis, M.V., Fotiadis, D.I. 2015. Machine learning applications in cancer prognosis and prediction. *Computational and Structural Biotechnology Journal*, 13: 8–17. http://dx.doi.org/10.1016/j.csbj.2014.11.005

Leatherdale, S.T., Lee, J. 2019. Artificial intelligence (AI) and cancer prevention: The potential application of AI in cancer control programming needs to be explored in population laboratories such as COMPASS. *Cancer Causes and Control*, 30: 671–675. https://doi.org/10.1007/s10552-019-01182-2

Levine A.B., Schlosser C., Grewal J., Coope R., Jones S.J.M., Yip S. (2019). Rise of the machines: Advances in deep learning for cancer diagnosis. *Trends in Cancer*, 5, 157–169.

NCI. General Information about Adult Brain Tumors. *NCI*. 14 April 2014. Available from: https://www.cancer.gov/types/brain/patient/adult-brain-treatment-pdq

Patil, S., Moafa, I.H., Alfaifi, M.M., Abdu, A.M., Jafer, M.A., Raju, K L., Raj, A.T., Sait, S.M. 2020. Reviewing the role of artificial intelligence in cancer. *Asian Pacific Jounal of Cancer Biology*, 5 (4): 189–199. DOI:10.31557/APJCB.2020.5.4.189

Ramya, V.J., Navarajan, J., Prathipa, R., Kumar, L.A. 2015. Detection of melanoma skin cancer using digital camera images. *ARPN Journal of Engineering and Applied Sciences*, 10: 3082–3085.

Ruddon, R.W. 2007. *Cancer Biology* (4th ed.). Oxford: Oxford University Press, p. 223.

Saba, T. 2020. Recent advancement in cancer detection using machine learning: Systematic survey of decades, comparisons and challenges, *Journal of Infection and Public Health*, 13: 1274–1289, https://doi.org/10.1016/j.jiph.2020.06.033

Selaru, F.M., Xu, Y., Yin, J., Zou, T., Liu, T.C., Mori, Y., Abraham, J.M., Sato, F., Wang, S., Twigg, C., Olaru, A., Shustova, V., Leytin, A., Hytiroglou, P., Shibata, D., Harpaz, N., Meltzer, S.J. 2002. Artificial neural networks distinguish among subtypes of neoplastic colorectal lesions. *Gastroenterology*, 122: 606–613. doi: 10.1053/gast.2002.31904.

Sherbet, G.V., Woo, W.L., Dlay, S. 2018. Application of artificial intelligence-based technology in cancer management: A commentary on the deployment of artificial neural networks. *Anticancer Research*, 38: 6607–6613. doi:10.21873/anticanres.13027 https://towardsdatascience.com/applied-deep-learning-part-1-artificial-neural-networks-d7834f67a4f6

Sobhaninia, Z., Rezaei, S., Noroozi, A., Ahmadi, M., Zarrabi, H., Karimi, N., Emami, A., Samavi, S. 2018. Brain tumor segmentation using deep learning by type specific sorting of images. Available at: https://arxiv.org/abs/1809.07786v1

Thenault, R., Kaulanjan, K., Darde, T., Rioux-Leclercq, N., Bensalah, K., Mermier, M., Khene, Z., Peyronnet, B., Shariat, S., Pradère, B., Mathieu, R. 2020. The application of artificial intelligence in prostate cancer management—what improvements can be expected? A systematic review. *Applied Sciences,* 10 (18): 6428. https://doi.org/10.3390/app10186428

Tibshirani, R., Hastie, T., Narasimhan, B., Chu, G. 2002. Diagnosis of multiple cancer types by shrunken centroids of gene expression. *Proceedings of the National Academy of Science*, 99: 6567–6577. https://doi.org/10.1073/pnas.082099299.

Wildeboer, R.R., van Sloun, R.J.G., Wijkstra, H., Mischi, M. 2020. Artificial intelligence in multi-parametric prostate cancer imaging with focus on deep-learning methods. *Computer Methods and Programs in Biomedicine,* 189: 105316. https://doi.org/10.1016/j.cmpb.2020.105316.

7 Detection of Diabetic Foot Ulcer Using Machine/Deep Learning

Dania Sadaf
Department of Computer Science, COMSATS University
Islamabad, Wah Campus, Wah Cantt, Pakistan

Javeria Amin
Department of Computer Science, University of Wah,
Wah Cantt, Pakistan

Muhammad Sharif
Department of Computer Science, COMSATS University
Islamabad, Wah Campus, Wah Cantt, Pakistan

Mussarat Yasmin
Department of Computer Science, COMSATS University
Islamabad, Wah Campus, Wah Cantt, Pakistan

7.1 INTRODUCTION

Diabetes is a disturbance in metabolism generated mainly by liver disease that in turn causes complications such as foot ulcer and retinopathy (Noor et al. 2015, Qureshi et al. 2016, Amin, Sharif, Rehman, et al. 2018). Throughout the world, at varying ages, approximately 2.8% of people were affected by diabetes in the year 2000 and this level will reach 4.4% by 2030 (Wild et al. 2004). In adults above 18 years, the global occurrence of diabetes was 8.5% in 2014 (Lipsky et al. 2016, World Health Organization 2016). For patients with diabetes, there is about 15–25% chance of eventually developing DFU; without proper care properly, DFU may result in foot or limb amputation (Aguiree et al. 2013). Type 1 diabetes is generated when insulin production is low, causing viral or bacterial infection and resulting in cell damage. In contrast, type 2 diabetes occurs as a result of an increase in blood glucose level because of resistance to insulin that damages organs of the body (Chimen et al. 2012).

7.1.1 Ulcer Disease and Types

The cause of ulcers is mainly untreated injury resulting from blood flow problems (Souaidi and El Ansari 2019). Usually, ulcer symptoms include swelling, itching, burn or rash, brown discoloration, dry skin, redness, and blisters (Ali et al. 2001). There are several types of ulcers, as follows:

1. *Peptic ulcers*: A peptic ulcer is a common type of stomach or duodenal ulcer. Usually, symptoms include epigastric pain (pain in the upper part of the stomach), vomiting, hematemesis, and melena. The most severe and frequent complications are obstruction, bleeding, perforation, and penetration (Chan and Leung 2002, Malfertheiner, Chan, and McColl 2009, Milosavljevic et al. 2011, Graham and Khalaf 2020).

2. *Esophageal ulcers*: Esophageal ulcers are also called gastric ulcers and develop in the stomach lining, inside the esophagus. They commonly arise after intake of food or when juices or stomach acid reflux into the esophagus and erode its lining (Tarnawski and Ahluwalia 2012).

3. *Duodenal ulcers*: Duodenal ulcers develop inside the duodenum (small intestine). A common symptom is epigastric pain that usually occurs during fasting or in the night and are usually relieved by taking food or agents that neutralize acid; some duodenal ulcers also provoke heartburn (Pounder 1981, Varacallo and Mair 2020).

4. *Helicobacter pylori ulcer*: *H. pylori* is a Gram-negative spirochete and is the main cause of stomach ulcers. It consists of an enzyme named urease which commonly alkalinizes the inflammatory part of the stomach or gastric mucosa and remains in the stomach for many years. Other causes include stress, alcohol, excessive consumption of caffeine, smoking, or genetics (Wu, Lin, and Liaw 1995, Malfertheiner, Chan, and McColl 2009).

5. *Arterial ulcer*: Arterial ulcers are also called ischemic ulcers; they featire on the outer side of the toes, ankles, hands, heels, feet, legs, etc. An arterial ulcer starts from a small, shallow wound that increases in depth and size with time. It is caused by decreased perfusion. The risk of fungal infection increases due to poor tissue oxygen concentration (Nelson and Bradley 2007, Bohnett et al. 2019).

6. *Venous ulcer*: This is a common type of leg, mouth, and foot ulcer. Generally, the shape of these ulcers is irregular and there is a moderate to high amount of wound drainage. A yellow thin fibrous coating covers the wound bed and looks glossy. It is often associated with many skin changes, i.e., dermatitis and dry, scaly skin (Bevis, Earnshaw, and Dermatology 2011, Evans et al. 2019).

7. *Diabetic foot ulcer*: DFU, also known as diabetic neuropathic ulcer, is the foremost problem of diabetes (Abbott et al. 1998, Paton et al. 2011). The cause of DFU may be ischemia that results in amputation of the foot if not treated or identified early (Santilli and Santilli 1999). DFU affects patients and especially those who are affected by gangrene or ischemia (Prompers et al. 2007). Ischemia is an inadequate supply of blood that can affect the healing of DFU, and it develops due to poor perfusion pressure to the foot (tissue damage of

that foot part) (Prompers et al. 2007). Infection is characterized by DFU bacterial soft-tissue infection on the basis of inflammation (Williams, Hilton, and Harding 2004). Bacterial infection, continuing trauma, tissue ischemia, and poor management cause DFU to heal slowly and it readily transforms into a chronic wound (Jeffcoate and Harding 2003). In the medical system for a diagnostic test, a blood test is performed as a diagnostic gold-standard test (Ince et al. 2008). It is necessary to develop robust methods using computer vision techniques for DFU detection (Irum et al. 2014) that include various grades and stages, based on the Texas classification. However, before a complete system of DFU diagnosis is developed, different conditions such as depth, site, area, ischemia, infection, and neuropathy should be considered with regard to the outcome of DFU (Lavery et al. 1996, Schaper and Reviews 2004, Treece et al. 2004, Jain 2015). Patients suffering from DFU – particularly those with ischemia – should be checked if they have an infection. Approximately 56% of people with DFU become infected, and 20% of DFU infections lead to limb or foot removal (Prompers et al. 2007, Lipsky et al. 2012).

7.1.2 REAL-TIME DFU APPLICATIONS

Automated smartphone applications for the recognition and detection of DFU diseases are becoming popular in image processing and computer vision. Yap et al. (2016, 2018) developed a FootSnap application that is used to produce a standardized DFU images dataset. Basic techniques for processing are used in this application, i.e., edge detection for producing foot ghost images useful in monitoring the progress of the DFU. Since it was designed to standardize conditions of image capture, it does not perform DFU automated recognition.

7.1.3 SCOPE AND OBJECTIVES

This chapter focuses on an overview of DFU, its challenges, and the methods utilized for its detection, as shown in Figure 7.1. This work considered different methods of image pre-processing, image segmentation, hand-crafted, and deep features extraction, features selection (Saba, Rehman, Jamail, et al. 2021, Saba, Rehman, Latif, et al. 2021), and classification, with their challenges.

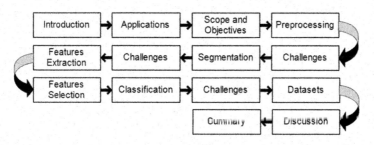

FIGURE 7.1 Organization of chapter.

7.2   PRE-PROCESSING METHODS

Pre-processing plays an essential role for noisy, inconsistent, and incomplete data and it is the initial step required to acquire high accuracy. The images captured by cameras contain a lot of noise and include poor background (Sharif et al. 2010, Jeyavathana et al. 2016).

Image enhancement is a process for adjusting digital images to make the results more suitable for image analysis. It comprises brightening, sharpening, and noise removal, which makes it easier to identify key features (Huang 1969, Irum et al. 2015). Data augmentation, i.e., rotating, mirror, Gaussian noise, contrast enhancement, translation, sharpen, and shear, was performed by Goyal et al. (2020) for performance enhancement. Then texture and color features were extracted and the proposed algorithm of superpixel color descriptor and deep ensemble architecture was applied for the classification of ischemia and infection.

Morphological operations, i.e., dilation, erosion, closing, and opening, were utilized by Hernandez-Contreras et al. (2015) for segmenting thermal patterns of diabetic foot ulcers followed by classification. Mwawado et al. (2020) performed resizing of images to convert them to double. Further, red, green, and blue (RGB) to grayscale conversion was performed to find edges and convert the edge map into a binary image. Morphological operation, i.e., closing, was applied to fill and connect edges of the ulcer area to segment DFU and introduce a robust edge detection method for DFU detection. Fraiwan, Ninan, and Al-Khodari (2018) converted images to grayscale for efficient processing then applied a binary threshold of 149 to segment patient feet from the background. The authors further utilized morphological operations of erosion for noise removal and dilation to recover foot area loss while mean temperature difference was used for foot ulcer and non-ulcer detection. In another study, image binarization was utilized by Solís-Sánchez et al. (2016) with a threshold of 254, after which morphological operations were applied, i.e., erosion for reduction of unwanted objects and a closing operation for the cleaning process used to eliminate isolated pixels of an image. In addition, segmentation of the DFU region was done. Another step of pre-processing is color space enhancement. This is a very basic and crucial step for contrast enhancement from the background and plays an essential role in the medical imaging field (Lakhwani, Murarka, and Chauhan 2015). Hue, saturation, value (HSV) color space was used by Godeiro et al. (2018) before segmentation and image binarization were performed to reduce noise and detect human skin and wounds. The segmented image was transformed into CIELAB color space to accelerate the classification procedure. Finally, wound tissue classification was done.

In pre-processing, another process is noise removal, used to reduce or remove noise from the noisy image to make the entire image smooth without losing information or details (Fan et al. 2019).

A method for noise filtering was developed by Chakraborty et al. (2016) using median filter, color correction using YDbDr color transformation, and color homogenization using anisotropic diffusion followed by segmentation of wound and tissue classification. In another study, a modified mean filter was applied by Cui et al. (2019) using a threshold to reduce the noise effect caused by light reflection. Then the diabetic lesion was extracted to apply diabetic wound segmentation. Also, post-processing was done to refine segmentation and obtain predicted masks. Patel, Patel, and Desai (2017) utilized RGB-to-HIS (hue, intensity, saturation) conversion and a diffusion procedure for noise removal from wound images as diffusion smoothes the wound tissues by preserving borders. After wound tissue segmentation and texture feature extraction, morphological operation filling was used to fill small holes, removing small regions and extracting the wound region to detect wound tissue.

7.2.1 CHALLENGES

The main challenges in this area were: (1) non-availability of public datasets and costly annotations to images of DFU; (2) DFU datasets lack standardization, such as camera distance from the patient's foot, brightness, lighting conditions, and image orientation; (3) performance degradation due to variations in shape, size, and texture; and (4) absence of meta-data-like patient age, ethnicity, sex, and size of the foot (Goyal 2019).

7.3 SEGMENTATION

Segmentation is an essential phase due to its property of simplifying the image representation and its analysis (Masood et al. 2015, Rodríguez-Esparza et al. 2020). Different techniques are discussed in this chapter, including region-based, semantic, instance, thresholding-based, color image, and deep learning-based segmentation (Khan et al. 2019, Nisa et al. 2020).

In the region-based segmentation method, region of interest (ROI) from an image is selected by examining neighbor pixels (Tang 2010). This involves dividing the images into uniform regions for processing based on an appropriate thresholding technique (Tremeau and Borel 1997, Kaganami and Beiji 2009). Adaptive rescaling is applied before segmentation (Song and Sacan 2012) to reduce time consumption. Then for automatic image segmentation of DFU images, wound regions are extracted using region growing, edge detection, and thresholding. Veredas, Mesa, and Morente (2009) implemented an algorithm of mean shift for region smoothing and a method of region growing for effective region segmentation. Another technique is semantic segmentation, also termed utilizing scene labeling. In this method a specific label is allocated to every image pixel (Yu et al. 2018).

In other words, each image pixel is classified into a class (an instance) (Taghanaki et al. 2021). An algorithm of mask region-based convolutional neural network (R-CNN) is trained by Muñoz et al. (2020) to carry out instance segmentation to discriminate lesions from images of DFU, achieving accuracy of 98.01%. These authors combined two important tasks of object detection for locating an object in an image and segmentation of natural instances. Clustering-based segmentation is used for partitioning image data into several clusters or disjoint groups (Sathya and Manavalan 2011). For differentiation of the ulcer region and its surroundings (Jawahar et al. 2020), RGB to HSV conversion is done, which makes human visualization easier. Dhane et al. (2017) used an order statistics filter for the removal of noise and redundant regions. Then they performed color normalization and RGB to YDbDr color transformation and presented fuzzy-spectral clustering (FSC) algorithm for wound region delineation, obtaining accuracy results of 91.5%. A system for wound analysis was proposed by Wang et al. (2014); this system runs on android smartphones. The mean shift algorithm was applied to the segmentation of wound images, achieving a Matthews correlation coefficient of 73.6%. In other research, accelerated mean shift was utilized by Bhelonde et al. (2015) for wound segmentation on android phones. Dalya and Shedge (2016) presented an algorithm of K-means shift for wound assessment based on smartphone and attained an accuracy of 80%. Moreover, edge-based segmentation is another technique for extracting edge information about images and locating areas with strong contrast, such as the object's location, size, and shape (Mageswari, Sridevi, and Mala 2013). Improved watershed segmentation technique was proposed by Babu, Ravi, and Sabut (2017) using an algorithm of pruning and flooding for diabetic wound-healing assessment. In the segmentation process, threshold-based methods are widely used to segment images and are useful for discriminating foreground from background (Kumar et al. 2014). Fraiwan et al. (2017) presented two techniques for performing analysis.

The segmentation process also includes color-based segmentation. It is based on color features of image pixels and each cluster defines the class of pixels containing similar color properties (Ganesan et al. 2019). RGB (a common representation of color) is aimed at storing images. Further colors could be achieved by RGB space non-linear or linear conversion. Liu et al. (2015) considered five color representations, i.e., RGB, RGB-ratio, normalized RGB, HSV, Y'CbCr. Kolesnik and Fexa (2005) proposed a technique of multidimensional color histogram sampling for feature descriptor and passed it as input to support vector machine (SVM) for automatic wound region extraction. Five techniques of histogram sampling comparison were performed:independent sampling (IS), learning vector quantization (LVQ), vector quantizer design (LBG-VQ), random density estimation (RDE), and multidimensional histogram sampling (HS). The problem of difficult wound segmentation can be generalized well by an SVM classifier using only 3D color histograms.

7.3.1 DEEP LEARNING-BASED SEGMENTATION

Deep learning is known to be a sub-field of machine learning (Amin, Sharif, Raza, et al. 2018, Amin, Sharif, Gul, et al. 2020, Sharif et al. 2021) and is also used for segmentation purposes. Bouallal et al. (2020) performed RGB to grayscale conversion and proposed a U-net (Ronneberger, Fischer, and Brox 2015) deep learning (Saba et al. 2020) model for plantar diabetic foot thermal image segmentation. In another study, U-net was utilized by Hernández et al. (2019). Ohura et al. (2019) utilized four architectures, namely Link-Net, Seg-Net, U-net, and U-net with Vgg16, for wound segmentation and detection. U-net outperformed the others, with an accuracy of 97.82% on the DFU dataset. Liu et al. (2017) presented WoundSeg architecture based on deep convolutional neural network (CNN) for wound segmentation and achieved 98.18% accuracy. Wang et al. (2015) cropped images using a modified version of GrabCut and introduced an architecture ConvNet based on encoder-decoder for wound area segmentation and analysis of the wound-healing process. Accuracy of 95.6% was acquired by their proposed technique. Maldonado et al. (2020) suggested a system that helps in detecting and classifying foot sole zone temperature differences, i.e., >2.2°C is ulcerous and <−2.2°C is necrotic. A retrained mask-RCNN model was used for segmenting images. Table 7.1 gives a summary of segmentation techniques.

TABLE 7.1
Summary of segmentation techniques

Authors	Methods	Datasets	Method type	Results
Bouallal et al. (2020)	RGB to grayscale conversion, proposed U-net architecture, dice coefficient	Private (394 thermal images)	Segmentation	Dice index of U-net with multimodal images = 99.17%, U-net with only thermal data = 98.68%
Ohura et al. (2019)	Link-Net, Seg-Net, U-net and U-net with Vgg16	Private (396 images)	Segmentation and detection	Accuracy = 97.82% on U-net, Sensitivity = 85.8%, specificity = 98.8%
Godeiro et al. (2018)	HSV color space, image binarization, watershed, GrabCut, Otsu thresholding, CIELAB color transformation, U-net	Private (30 images)	Segmentation and classification	Accuracy by U-net = 96.10%, dice coefficient = 94.25%, sensitivity = 91.28%, specificity = 98.76%
Goyal et al. (2017)	Semantic segmentation by their proposed FCN, dice coefficient	Private (705 images)	Segmentation	Dice-index = 89.9% of surrounding region, and 79.4% ulcer region
Babu, Ravi, and Sabut (2017)	Proposed an improved watershed algorithm using pruning and flooding	Private	Segmentation	Tested three images: 99.24% on image 1, 100% on image 2, and 97.85% on image 3
Etehadtavakol et al. (2017)	Snakes algorithm (active contour algorithm), asymmetry analysis, fuzzy c-mean	Private (59 images)	Segmentation	Temperature > 3°C for ulcer detection
Dalya and Shedge (2016)	K-mean shift algorithm	Private	Segmentation	Accuracy = 80%
Liu et al. (2015)	RGB, RGB ratio, normalized RGB, HSV, Y'CbCr, and CIE L*a*b* and k-mean	Private (76 patients' data)	Segmentation	Specificity = 99%, and sensitivity = 98% in CIEL*a*b*

(continued)

D. Sadaf et al.

TABLE 7.1 Cont.
Summary of segmentation techniques

Authors	Methods	Datasets	Method type	Results
Wang et al. (2014)	Downsample images by factor 4 in both vertical and horizontal direction, Gaussian blur method using SD = 0.5 to remove noise, color segmentation based on K-mean algorithm, mean shift algorithm	Private (64 images)	Segmentation	MCC = 73.6%
Song and Sacan (2012)	Adaptive rescaling, eegion growing, edge detection, and thresholding	Private (19 patients' data)	Segmentation	Accuracy by RBF = 85.7%, MLP = 71.4%
Veredas, Mesa, and Morente (2009)	Mean shift algorithm, region growing, multilayer perceptron	Private (113 images)	Segmentation and classification	Accuracy = 91.5%, specificity = 94.7%, sensitivity = 78.7%
Kolesnik and Fexa (2006)	Proposed SVM-based segmentation	Private	Segmentation	Output segmentation quality varies according to error rate, ranging from 0.47% to 30.45%

RGB, red, green, and blue; HSV, hue, saturation, value; FCN, fully convolutional network; MCC, Matthews correlation coefficient; RBF, radial basis function; MLP, multiple-layer perceptron; SVM, support vector machine.

7.3.2 CHALLENGES

In image segmentation, interactive techniques provide appropriate results but as the amount of images increases, these techniques become non-feasible. For automatic segmentation of images, many techniques have been introduced along with several challenges that involve low contrast, noisy images, and high computation time (Fida et al. 2017). Furthermore, expensive imaging modalities like magnetic resonance inductance and X-ray may not be easily accessed in many developing countries that have limited healthcare facilities. These challenges require the development of methods to objectify and enhance accuracy to analyze DFU using traditional methods (Lepäntalo et al. 2011).

7.4 FEATURES EXTRACTION

Features extraction is an influential phase in image processing used in the extraction of useful features for classification and image recognition (Haider et al. 2014). Common features are extracted by the process of features extraction (Pachouri and Technologies 2015). Two types of features are hand-crafted and deep features (Saba et al. 2019).

7.4.1 HAND-CRAFTED FEATURE-BASED METHODOLOGIES

Hand-crafted features are utilized to extract information (features) from images (Nanni, Ghidoni, and Brahnam 2017). Over the years, several feature descriptors have been introduced. Different methods of handcrafted features extraction are discussed in this chapter, including texture and color-based features. In image processing, textural features are used for object recognition and classification. There are several methods for texture feature extraction. Goyal, Reeves, Davison et al. (2018) performed data augmentation and then extracted features using histogram orientation gradient (HOG), local binary pattern (LBP), color descriptors (HSV, RGB, Lu*v*), and trained sequential minimal optimization (SMO), and also proposed DFUNet for normal/abnormal (ulcer) classification. Saminathan et al. (2020) segmented right and left foot regions and extracted 12 gray-level co-occurrence matrix (GLCM) features (correlation, autocorrelation, dissimilarity, contrast, entropy, energy, maximum probability, homogeneity, sum entropy, variance, sum variance, and sum average) along with two temperature features (mean and standard deviation of temperature). These extracted features were classified for computer-aided detection of DFU (Wu et al. 2000). Then the best features were selected and a two-stage SVM-based classifier was developed for DFU area determination. In another study, texture, color, and features were extracted by Mukherjee et al. (2014). The HSI, Luv, Lab, HSL, HSV, YIQ, YUV, CAT02 LMS, LCH, Y'CbCr, JPEG-Y'CbCr, YDbDr, and YPbPr were taken for color features considering 45 color channels. Extraction of five features based on color, i.e., mean, variance, standard deviation, kurtosis, and skewness, was done from each of 45 color channels for every ROI along with the extraction of ten textural features. Further, these texture and color features were provided to classifiers for automated wound tissue classification. In features extraction, wavelet-based features are also used (Erişti, Uçar, and Demir 2010, Adam et al. 2018). DFU detection and analysis were performed by Vali, Sharma, and Ahmed (2017) using a modified chan vase algorithm in which firstly resizing of images was done followed by gray to binary conversion. Then distance transform was used to extract geometric features. Lastly, global region-based segmentation was performed.

7.4.2 DEEP FEATURES-BASED METHODOLOGIES

A deep architecture (Litjens et al. 2017, Raza et al. 2018, Voulodimos et al. 2018, Amin et al. 2019, Fayyaz et al. 2020, Naz et al. 2021), DFU_QUTNet, was proposed by Alzubaidi et al. (2020) for the extraction of deep features; AlexNet, Googlenet, and Vgg16 were used for comparison purposes. Then DFU classification was performed using SVM and K-nearest neighbor (KNN), achieving the highest F1-score of 95.61% on SVM. Goyal, Reeves, Rajbhandari, et al. (2018) utilized SSD-mobilenet, SSD-inceptionv2, Resnet101 with RFCN, and Inceptionv2 with R-CNN for real-time DFU detection and localization, as shown in Figure 7.2.

In another study, Goyal and Hassanpour 2020) firstly applied data augmentation and then performed experimentation using faster R-CNN, Yolov5, and proposed a computer-aided detection (CAD) system with EfficientDet architecture for DFU detection, acquiring accuracy ranging from 98.4% to 99.9%. Ensemble features are also famous for better performance. Wijesinghe et al. (2019) applied mask-RCNN and proposed deep-ensemble CNN for features extraction, after which they used singular value decomposition (SVD) and artificial neural networks (ANN) for wound classification. Amin, Sharif, Anjum, et al. (2020) extracted features using their proposed CNN architecture and utilized different classifiers for the classification of ischemia and infection, as shown in Figure 7.3. The highest accuracy of 97.9% was achieved on naive Bayes for ischemia detection and 99.6% on the decision tree for infection detection. Further, for the localization of the infected region, the YOLOv2-DFU network is used. Han et al. (2020) utilized SSD-mobilev2, Faster RCNN-Resnet101 models, and YOLOv3 refinements for real-time detection and diabetic foot determination. Greater accuracy (91.95%) was attained after refinement of YOLO-v3. Table 7.2 presents a summary of deep features extraction methods.

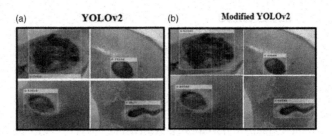

FIGURE 7.2 Localization of diabetic foot ulcer: (a) YOLOv2; (b) modified YOLOv2 (Amin, Sharif, Anjum, et al. 2020).

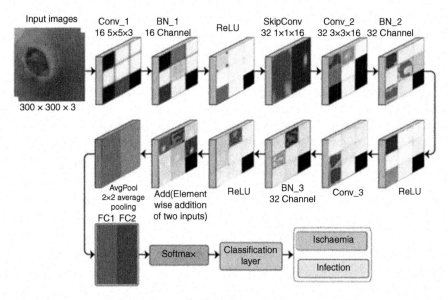

FIGURE 7.3 Classification of ischemia and infection using convolutional neural network (Amin, Sharif, Anjum, et al. 2020).

TABLE 7.2
Summary of deep features extraction methods

Author	Method	Dataset	Method type	Results
Goyal and Hassanpour (2020)	R-CNN, Yolov5, and proposed CAD system using EfficientDet architecture	DFUC dataset (4500 images)	Detection	Accuracy ranging from 98.4% to 99.9%
Alzubaidi et al. (2019)	Proposed DFU_QUTNet, AlexNet, Googlenet, Vgg16, SVM, KNN	Private (754 images)	Classification	F1-score = 94.5% using SVM
Goyal, Reeves, Rajbhandari, et al. (2018)	SSD-mobilenet, SSD-inceptionv2, Resnet101 with RFCN, and Inceptionv2 with faster R-CNN	Private	Detection and localization	Accuracy = 91.8%

R-CNN, region-based convolutional neural network; CAD, computer-aided detection; SVM, support vector machine.

7.5 FEATURES SELECTION

In features selection, the best features are selected by removing irrelevant and redundant data from original features. This is also known as feature reduction, feature optimization, and dimensionality reduction (Khan et al. 2020). Sivayamini et al. (2017) extracted texture features to apply the cuckoo search optimization technique for the collection of finest features from the extracted features. Kaabouch et al. (2011) proposed an enhanced algorithm for foot inflammation detection and ulcer prediction. Based on asymmetry analysis, this algorithm was combined with the technique of segmentation using a genetic algorithm to achieve greater efficiency in inflammation detection. The genetic algorithm effectively crops feet from the background and eliminates most noise.

7.6 CLASSIFICATION

Classification is performed to categorize images into the classes to which they belong. Babu, Sabut, and Nithya (2018) performed segmentation to extract ROI and then extracted texture features. Afterwards, they utilized two classifiers, named Hoeffeding tree classifier and naive Bayes classifier, on the extracted DFU features, and achieved high accuracy of 90.0% on naive Bayes. Sudarvizhi et al. (2019) used SVM as a classifier in their research. They used the load cell technique that includes image segmentation to identify and analyze foot ulcers by foot pressure analysis. In another work, SVM was used by Wannous et al. (2008) for the classification of wound tissue. A prediction system was proposed by Botros et al. (2016) for DFU. Wannous, Lucas, and Treuillet (2010) applied color and texture feature extraction and fed the extracted features to SVM, KNN, fuzzy-KNN, and K-means clustering for wound tissue classification, acquiring a kappa coefficient of 81% on SVM. Yusuf et al. (2015) suggested single microbial species and poly-microbial species classification in vitro for targeting infection of diabetic foot. Then SVM, KNN, linear discriminant analysis (LDA), and probabilistic neural network (PNN) were used for classification. Vardasca et al. (2019) analyzed collected data using the WEKA analysis framework (Hall et al. 2009) and implemented ANN, KNN, and SVM for monitoring DFU (neuro-ischemic and ischemic), achieving best results with KNN, i.e., 81.25% accuracy, 100% sensitivity, and 80% specificity. CNNs are a class of neural networks that has been determined to be very compelling in areas like classification and image recognition. Badea et al. (2016) introduced a classification method by which burn wounds could be distinguished from healthy skin based on CNN network, i.e., MatConNet architecture, along with other methods including deep learned RGB method, deep learned scaled RGB method, hue saturation method, color texture method, and skin model (TGB-Yuv) method.

7.6.1 CHALLENGES

Since DFU analysis utilizing computerized methods is a comparatively developing field, it consists of a limited amount of computer methods for diabetic foot disease assessment developed using simple image processing and traditional methods of machine learning (Liu et al. 2015, Wang et al. 2016). High similarity among classes and variations within a class rely on different DFU classification techniques (Guo et al. 2020). In Table 7.3, a summary of different methods used for classification and detection is given along with their results.

TABLE 7.3
Summary of classification techniques

Author	Method	Dataset	Method type	Results
Saminathan et al. (2020)	GLCM features, mean and SD of temperature features, SVM	Private	Detection	Accuracy = 95.61%
Sudarvizhi et al. (2019)	Histogram-based segmentation, otsu-thresholding, point-to-point mean difference, SVM	Private	Foot pressure analysis, segmentation, classification	Accuracy = 96.4%
Fraiwan, Ninan, and Al-Khodari (2018)	Mobile application using grayscale transformation, erosion, dilation, image binarization using threshold 149, MTD	Private	Detection	If MTD > 2.2 ulcer detected, if MTD < 2.2 non-ulcer
Babu, Sabut, and Nithya (2018)	GLCM, Ho-effeding tree, and naive Bayes classifier	Private	Classification	Accuracy = 90% on naive Bayes
Sivayamini et al. (2017)	Median filter, DWT, GLCM, PSO, CS	Private	Detection	Accuracy = 79.109% on CS, 28.57% on PSO
Botros et al. (2016)	WEKA data-mining software, SVM	Private	Classification	Accuracy = 96.4%
Yusuf et al. (2015)	E-nose technique, SVM, KNN, LDA, and PNN	Private	Classification	Accuracy = 90%
Kaabouch et al. (2011)	Geometry transformation algorithm, mean, variance, skewness, kurtosis, entropy, joint entropy, correlation, Otsu method, genetic algorithm	Private	Detection	Out of 100 images, 55 were recognized without hot spots, 45 with hot spots, 10 with false spots, and 0 with missing spots

(continued)

TABLE 7.3 Cont.
Summary of classification techniques

Author	Method	Dataset	Method type	Results
(Wannous, Lucas, and Treuillet 2010)	MCD and DCD color descriptors, GLCM, SVM	Private	Classification	Accuracy = 81%
Wannous et al. (2008)	JSEG algorithm, SVM	Private	Segmentation, classification, and detection	Mean accuracy on 2D images 77.5%, 69.4%, 69.8%, on 3D images 80.6%, 69.8%, 71.5%
Wannous, Lucas, and Treuillet (2008)	CSC, JSEG, mean shift, graph segmentation methods, color features (RGB and CIELa*b*), LBP, texture contrast, anisotropy, proposed efficient SVM	Private	Classification	Classification overlap score (66–88%)

GLCM, gray-level co-occurrence matrix; SD, standard deviation; SVM, support vector machine; MTD, mean temperature difference; DWT, discrete wavelet transform; PSO, particle swarm optimization; CS, cuckoo search optimization; KNN, K-nearest neighbor; LDA, linear discriminant analysis; PNN, probabilistic neural network; MCD, methionine- and choline-deficient; DCD, dominant color descriptor; JSEG, J-based segmentation; 2D, two-dimensional; CSC, convolutional sparse coding; RGB, red, green, and blue; LBP, local binary pattern.

7.7 DATASETS

Most DFU datasets are not publicly available and data collection is a major challenge (Goyal et al. 2020). DFU datasets that are publicly available include Medetec Wound Database (Thomas), proposed mainly for DFU wound area segmentation (Jawahar et al. 2020), DFU segmentation, and wound tissue classification (Patel, Patel, and Desai 2017). Another public dataset, NYU database, is proposed for wound segmentation (Wang et al. 2015). Mostly there are private datasets, some of which are described here. In Goyal et al. (2020) and Goyal, Reeves, Davison, et al. (2018), the researchers introduced their own DFU dataset (Yap) of 397 normal/abnormal (ulcer) images and 1,459 images for ischemia and infection. They then applied data augmentation to increase the dataset. They made the dataset available upon request. They collected the dataset from Lancashire Teaching Hospital and obtained approval from the UK National Health Service research ethics committee. In Goyal and Hassanpour (2020), the authors collected the DFUC2020 dataset of 4,500 images of DFU and performed DFU detection using 90% data for training and 10% for testing. In the DFU dataset, there were 1,459 ischemia (negative/positive) images and 1,459 images of infection (negative/positive). Further, the researchers applied natural data augmentation and their dataset was enhanced to 9,870 augmented ischemia patches and 5,892 augmented infection patches.

7.8 DISCUSSION

The major DFU challenges affecting accuracy are brightness, lighting conditions, noise, and image orientation. Other challenges include a variety of shapes, sizes, and textures that affect the performance of classification. Further, unbalanced class distribution causes difficulties in accurate DFU detection. In addition, methods discussed in this chapter include preprocessing, data augmentation, morphological operations, image resizing, histogram equalization, noise removal, and color transformations. In DFU segmentation, region-based, semantic, and instance segmentation, and segmentation by deep learning have been discussed. In DFU features extraction, different types of texture, color, HOG, wavelet, geometric, entropy, and deep CNN features have been described. Likewise, in DFU features selection, methods discussed included SVD, linear programming problems (LPP), non-paternity event (NPE), principal component analysis (PCA), Learning and Skills Development Agency (LSDA), kernel principal component analysis (KPCA), and genetic algorithm. In the classification of DFU, different classifiers were considered, i.e., SVM, KNN, naive Bayes, PNN, Ho-effeding tree, random forest, fuzzy-KNN, and deep CNN.

7.9 SUMMARY

This chapter gives a review on DFU in which the task of detecting DFU is challenging due to dataset complexities affecting accuracy, such as brightness, lighting conditions, and image orientation. Other limitations and challenges based on the literature include noise that appears in DFU images during image acquisition, such as illumination and blurriness that affect the classification accuracy, variability in shape, size, and texture. Unbalanced class distribution causes difficulties in accurate DFU detection; similarities among different classes and dissimilarities within a class in DFU between different cases make it challenging for classifiers to predict the class accurately. In this chapter, challenges that occurred during preprocessing, segmentation, and classification that are not discussed in the existing literature were also highlighted.

REFERENCES

Abbott, Caroline A, Loretta Vileikyte, Sheila Williamson, Anne L Carrington, and Andrew JMJ Boulton. "Multicenter study of the incidence of and predictive risk factors for diabetic neuropathic foot ulceration," *Diabetes Care.* 1998. 21(7):1071–1075.

Adam, Muhammad, Eddie YK Ng, Shu Lih Oh, Marabelle L Heng, Yuki Hagiwara, Jen Hong Tan, Jasper WK Tong, U Rajendra, and Acharya. "Automated characterization of diabetic foot using nonlinear features extracted from thermograms." *Infrared Physics and Technology.* 2018. 89:325–337.

Aguiree, Florencia, Alex Brown, Nam Ho Cho, Gisela Dahlquist, Sheree Dodd, Trisha Dunning, Michael Hirst, Christopher Hwang, Dianna Magliano, Chris Patterson,

Courtney Scott, Jonathon Shaw, Guyula Soltesz, Juliet Usher-Smith, and David Whiting 2013, *IDF Diabetes Atlas*, 6th ed. Edited by Guariguata, Leonor, Nolan, Tim, Beagley, Jessica, Linnenkamp, Ute and Jacqmain, Olivier, International Diabetes Federation, Basel, Switzerland.

Ali, S M, A Basit, T Sheikh, S Mumtaz, and M Z I Hydrie. "Diabetic foot ulcer – a prospective study." *Journal-PAKISTAN Medical Association*. 2001. 51(2):78–80.

Alzubaidi, Laith, Mohammed A Fadhel, Sameer R Oleiwi, Omran Al-Shamma, and Jinglan J Zhang. "DFU_QUTNet: diabetic foot ulcer classification using novel deep convolutional neural network." *Multimedia Tools and Applications*. 2020. 79(21):15655–15677.

Amin, Javeria, Muhammad Sharif, Mudassar Raza, Tanzila Saba, and Muhammad Almas Anjum. "Brain tumor detection using statistical and machine learning method." *Computer Methods and Programs in Biomedicine*. 2019. 177:69–79.

Amin, Javeria, Muhammad Sharif, Muhammad Almas Anjum, Habib Ullah Khan, Muhammad Sheraz Arshad Malik, and Seifedine Kadry. "An integrated design for classification and localization of diabetic foot ulcer based on CNN and YOLOv2-DFU models." *IEEE Access*. 2020. 8:228586–228597.

Amin, Javeria, Abida Sharif, Nadia Gul, Muhammad Almas Anjum, Muhammad Wasif Nisar, Faisal Azam, and Syed Ahmad Chan Bukhari. "Integrated design of deep features fusion for localization and classification of skin cancer." *Pattern Recognition Letters*. 2020. 131:63–70.

Amin, Javeria, Muhammad Sharif, Amjad Rehman, Mudassar Raza, and Muhammad Rafiq Mufti. "Diabetic retinopathy detection and classification using hybrid feature set." *Microscopy Research and Technique*. 2018. 81(9):990–996.

Amin, Javeria, Muhammad Sharif, Mudassar Raza, and Mussarat Yasmin. "Detection of brain tumor based on features fusion and machine learning." *Journal of Ambient Intelligence and Humanized Computing*. 2018. https://doi.org/10.1007/s12652-018-1092-9.

Babu, K S, Kumar Y B Ravi, and Sukanta Sabut. "An improved watershed segmentation by flooding and pruning algorithm for assessment of diabetic wound healing." 2nd IEEE International Conference on Recent Trends in Electronics, Information and Communication Technology (RTEICT). 2017. 679–683.

Babu, K S, Sukanta Sabut, and D K Nithya. "Efficient detection and classification of diabetic foot ulcer tissue using PSO technique." *International Journal of Engineering Technology*. 2018. 7(3):1006–1010.

Badea, Mihai-Sorin, Constantin Vertan, Corneliu Florea, Laura Florea, and Silviu Bădoiu. "Automatic burn area identification in color images." 2016 International Conference on Communications (COMM). 2016. 65–68.

Bevis, Paul, and Jonothan Earnshaw. "Venous ulcer review." *Clinical, Cosmetic and Investigational Dermatology*. 2011. 4:7–14.

Bhelonde, Ashish, Nikhil Didolkar, Shubham Jangale, and Nikhil L Kulkarni. "Flexible wound assessment system for diabetic patient using android smartphone." *International Conference on Green Computing and Internet of Things*. 2015. 466–469.

Bohnett, Mary Clare, Michael Heath, Stephanie Mengden, and Lynne Morrison. "Arterial hand ulcer: A common disease in an uncommon location." *JAAD Case Reports*. 2019. 5(2):147–149.

Botros, Fady S, Mona F Taher, Naglaa M ElSayed, and Ahmed S Fahmy. "Prediction of diabetic foot ulceration using spatial and temporal dynamic plantar pressure." In 8th Cairo International Biomedical Engineering Conference. 2016. 43–47.

Bouallal, Doha, Asma Bougrine, Hassan Douzi, Rachid Harba, Raphael Canals, Luis Vilcahuaman, and Hugo Arbanil. "Segmentation of plantar foot thermal images: application to diabetic foot diagnosis." In 2020 International Conference on Systems, Signals and Image Processing. 2020. 116–121.

Chakraborty, Chinmay, Bharat Gupta, Soumya K Ghosh, Dev K Das, and Chandan Chakraborty. "Telemedicine supported chronic wound tissue prediction using classification approaches." *Journal of Medical Systems*. 2016. 40(3):1–12.

Chan, Francis K L, and W K Leung. "Peptic-ulcer disease." *The Lancet*. 2002. 360(9337): 933–941.

himen, Myriam, Amy Kennedy, Krishnarajah Nirantharakumar, T T Pang, R Andrews, and Partheepan Narendran. "What are the health benefits of physical activity in type 1 diabetes mellitus? A literature review." *Diabetologia*. 2012. 55(3):542–551.

Cui, Can, Karl Thurnhofer-Hemsi, Reza Soroushmehr, Abinash Mishra, Jonathan Gryak, Enrique Domínguez, Kayvan Najarian, and Ezequiel López-Rubio. "Diabetic wound segmentation using convolutional neural networks." In 2019 41st Annual International Conference of the IEEE Engineering in Medicine and Biology Society. 2019. 1002–1005.

Dalya, Vidyashree, and D K Shedge. "Design of smartphone-based wound assessment system." International Conference on Automatic Control and Dynamic Optimization Techniques. 2016. 709–712.

Dhane, Dhiraj Manohar, Maitreya Maity, Tushar Mungle, Chittaranjan Bar, Arun Achar, Maheshkumar Kolekar, and Chandan Chakraborty. "Fuzzy spectral clustering for automated delineation of chronic wound region using digital images." *Computers in Biology and Medicine*. 2017. 89:551–560.

Dick, Florian, J-B Ricco, A H Davies, P Cao, C Setacci, G de Donato, François Becker et al. "Chapter VI: follow-up after revascularisation." *European Journal of Vascular and Endovascular Surgery*. 2011. 42:575–590.

Erişti, Hüseyin, Ayşegül Uçar, and Yakup Demir. "Wavelet-based feature extraction and selection for classification of power system disturbances using support vector machines." *Electric Power Systems Research*. 2010. 80(7):743–752.

Evans, Robyn, Janet L Kuhnke, Cathy Burrows, Ahmed Kayssi, Chantal Labrecque, Deirdre O'Sullivan-Drombolis, and Pamela Houghton. "Prevention and management of venous leg ulcers." *Wounds Canada*. 2019. 1–70.

Fan, Linwei, Fan Zhang, Hui Fan, and Caiming Zhang. "Brief review of image denoising techniques." *Visual Computing for Industry, Biomedicine, and Art*. 2019. 2(1):1–12.

Fayyaz, Muhammad, Mussarat Yasmin, Muhammad Sharif, Jamal Hussain Shah, Mudassar Raza, and Tassawar Iqbal. "Person re-identification with features-based clustering and deep features." *Neural Computing and Applications*. 2020. 32(14):10519–10540.

Fida, Erum, Junaid Baber, Maheen Bakhtyar, Rabia Fida, and Muhammad Javid Iqbal. "Unsupervised image segmentation using lab color space." In Intelligent Systems Conference (IntelliSys). 2017. 774–778.

Fraiwan, Luay, Jolu Ninan, and Mohanad Al-Khodari. "Mobile application for ulcer detection." *The Open Biomedical Engineering Journal*. 2018. 12:16–26.

Fraiwan, Luay, Mohanad AlKhodari, Jolu Ninan, Basil Mustafa, Adel Saleh, and Mohammed Ghazal. "Diabetic foot ulcer mobile detection system using smart phone thermal camera: A feasibility study." *Biomedical Engineering*. 2017. 16 (1):1–19.

Ganesan, P, B S Sathish, K Vasanth, V G Sivakumar, M Vadivel, and C N Ravi. "A comprehensive review of the impact of color space on image segmentation." In 5th International Conference on Advanced Computing and Communication Systems. 2019. 962–967.

Godeiro, Vitor, José Silva Neto, Bruno Carvalho, Bruno Santana, Julianny Ferraz, and Renata Gama. "Chronic wound tissue classification using convolutional networks and color space reduction." IEEE 28th International Workshop on Machine Learning for Signal Processing. 2018. 1–6.

Goyal, Manu, Neil D Reeves, Adrian K Davison, Satyan Rajbhandari, Jennifer Spragg, and Moi Hoon Yap. "Dfunet: Convolutional neural networks for diabetic foot ulcer classification." *IEEE Transactions on Emerging Topics in Computational Intelligence.* 2018. 4(5):728–739.

Goyal, Manu, Neil D Reeves, Satyan Rajbhandari, and Moi Hoon Yap. "Robust methods for real-time diabetic foot ulcer detection and localization on mobile devices." *IEEE Journal of Biomedical and Health Informatics.* 2018. 23(4):1730–1741.

Goyal, Manu, Neil D Reeves, Satyan Rajbhandari, Naseer Ahmad, Chuan Wang, and Moi Hoon Yap. "Recognition of ischaemia and infection in diabetic foot ulcers: Dataset and techniques." *Computers in Biology and Medicine.* 2020. 117:1–10.

Goyal, Manu. "Novel Computerised Techniques for Recognition and Analysis of Diabetic Foot Ulcers." PhD diss., Manchester Metropolitan University, 2019.

Graham, David Y, and Natalia Khalaf. "Peptic ulcer disease." *Geriatric Gastroenterology.* 2020: 1–31.

Guo, Linjie, Hui Gong, Qiushi Wang, Qiongying Zhang, Huan Tong, Jing Li, Xiang Lei et al. "Detection of multiple lesions of gastrointestinal tract for endoscopy using artificial intelligence model: A pilot study." *Surgical Endoscopy.* 2020. 35(12):6532–6538.

Haider, Waqas, Hadia Bashir, Abida Sharif, Irfan Sharif, and Abdul Wahab. "A survey on face detection and recognition approaches." *Research Journal of Recent Sciences.* 2014. 1–8.

Hall, Mark, Eibe Frank, Geoffrey Holmes, Bernhard Pfahringer, Peter Reutemann, and Ian H Witten. "The WEKA data mining software: an update." *ACM SIGKDD Explorations Newsletter.* 2009. 11(1):10–18.

Han, Aifu, Yongze Zhang, Ajuan Li, Changjin Li, Fengying Zhao, Qiujie Dong, Qin Liu et al. "Efficient refinements on YOLOv3 for real-time detection and assessment of diabetic foot Wagner grades." arXiv preprint, arXiv:2006.02322. 2020. 1–11.

Hernández, Abián, Natalia Arteaga-Marrero, Enrique Villa, Himar Fabelo, Gustavo M. Callicó, and Juan Ruiz-Alzola. "Automatic segmentation based on deep learning techniques for diabetic foot monitoring through multimodal images." International Conference on Image Analysis and Processing. Springer, Cham. 2019. 414–424.

Hernandez-Contreras, D, H Peregrina-Barreto, J Rangel-Magdaleno, J Ramirez-Cortes, and F Renero-Carrillo. "Automatic classification of thermal patterns in diabetic foot based on morphological pattern spectrum." *Infrared Physics and Technology.* 2015. 73:149–157.

Huang, T S "Image enhancement: A review." *Opto-electronics.* 1969. 1(1):49–59.

Ince, Paul, Zulfiqarali G Abbas, Janet K Lutale, Abdul Basit, Syed Mansoor Ali, Farooq Chohan, Stephan Morbach, Jörg Möllenberg, Fran L Game, and William J. Jeffcoate. "Use of the SINBAD classification system and score in comparing outcome of foot ulcer management on three continents." *Diabetes Care.* 2008. 31(5):964–967.

Irum, I, M A Shahid, M Sharif, and M Raza. "A review of image denoising methods." *Journal of Engineering Science and Technology Review.* 2015. 8(5):1–10.

Irum, Isma, Muhammad Sharif, Mussarat Yasmin, Mudassar Raza, and Faisal Azam. "A noise adaptive approach to impulse noise detection and reduction." *Nepal Journal of Science and Technology.* 2014. 15(1):67–76.

Jain, Amit Kumar C. "A simple new classification for diabetic foot ulcers." *Medicine Science.* 2015. 4(2):2109–2120.

Jawahar, Malathy, L Jani Anbarasi, S Graceline Jasmine, and Modigari Narendra. "Diabetic foot ulcer segmentation using color space models." In 2020 5th International Conference on Communication and Electronics Systems. 2020. 742–747.

Jeffcoate, William J, and Keith G Harding. "Diabetic foot ulcers." *The Lancet.* 2003. 361(9368):1545–1551.

Jeyavathana, R Beaulah, R Balasubramanian, and A Anbarasa Pandian. "A survey: analysis on pre-processing and segmentation techniques for medical images." *International Journal of Research and Scientific Innovation.* 2016. 3:1–10.

Kaabouch, Naima, Yi Chen, Wen-Chen Hu, Julie W Anderson, Forrest Ames, and Rolf Paulson. "Enhancement of the asymmetry-based overlapping analysis through features extraction." *Journal of Electronic Imaging.* 2011. 20(1):1–8.

Kaganami, Hassana Grema, and Zou Beiji. "Region-based segmentation versus edge detection." In 2009 Fifth International Conference on Intelligent Information Hiding and Multimedia Signal Processing. 2009. 1217–1221.

Khan, Muhammad Attique, Muhammad Imran Sharif, Mudassar Raza, Almas Anjum, Tanzila Saba, and Shafqat Ali Shad. "Skin lesion segmentation and classification: A unified framework of deep neural network features fusion and selection." *Expert Systems.* 2019. 1–21.

Khan, Muhammad Attique, Tallha Akram, Muhammad Sharif, Kashif Javed, Mudassar Raza, and Tanzila Saba. "An automated system for cucumber leaf diseased spot detection and classification using improved saliency method and deep features selection." *Multimedia Tools and Applications.* 2020. 79(25):18627–18656.

Kolesnik, Marina, and Ales Fexa. "Multi-dimensional color histograms for segmentation of wounds in images." International Conference Image Analysis and Recognition, Springer, Berlin, Heidelberg. 2005. 1014–1022.

Kumar, M Jogendra, G V S Raj Kumar, and R Vijay Kumar Reddy. "Review on image segmentation techniques." *International Journal of Scientific Research Engineering and Technology.* 2014. 2278–0882.

Lakhwani, Kamlesh, P D Murarka, and N S Chauhan. "Color space transformation for visual enhancement of noisy color image." *International Journal of ICT Management.* 2015. 3(2): 9–11.

Lavery, Lawrence A, David G Armstrong, and Lawrence B Harkless. "Classification of diabetic foot wounds." *The Journal of Foot and Ankle Surgery.* 1996. 35(6): 528–531.

Lipsky, Benjamin A, Anthony R Berendt, Paul B Cornia, James C Pile, Edgar J G Peters, David G Armstrong, H Gunner Deery et al. "2012 Infectious Diseases Society of America clinical practice guideline for the diagnosis and treatment of diabetic foot infections." *Clinical Infectious Diseases.* 2012. 54 (12): 132–173.

Lipsky, Benjamin A, Javier Aragón-Sánchez, Mathew Diggle, John Embil, Shigeo Kono, Lawrence Lavery, Éric Senneville et al. "IWGDF guidance on the diagnosis and management of foot infections in persons with diabetes." *Diabetes/Metabolism Research and Reviews.* 2016. 32: 45–74.

Litjens, Geert, Thijs Kooi, Babak Ehteshami Bejnordi, Arnaud Arindra Adiyoso Setio, Francesco Ciompi, Mohsen Ghafoorian, Jeroen Awm Van Der Laak, Bram Van Ginneken, and Clara I. Sánchez. "A survey on deep learning in medical image analysis." *Medical Image Analysis.* 2017. 42: 60–88.

Liu, Chanjuan, Jaap J van Netten, Jeff G Van Baal, Sicco A Bus, and Ferdi van Der Heijden. "Automatic detection of diabetic foot complications with infrared thermography by asymmetric analysis." *Journal of Biomedical Optics.* 2015. 20 (2):1–11.

Liu, Xiaohui, Changjian Wang, Fangzhao Li, Xiang Zhao, En Zhu, and Yuxing Peng. "A framework of wound segmentation based on deep convolutional networks." In 2017 10th International Congress on Image and Signal Processing, BioMedical Engineering and Informatics (CISP-BMEI). 2017. 1–7.

Maldonado, H, R Bayareh, I A Torres, A Vera, J Gutiérrez, and L Leija. "Automatic detection of risk zones in diabetic foot soles by processing thermographic images taken in an uncontrolled environment." *Infrared Physics and Technology.* 2020. 105: 1–21.

Malfertheiner, Peter, Francis K L Chan, and Kenneth E L McColl. "Peptic ulcer disease." *The Lancet.* 2009. 374(9699):1449–1461.

Masood, Saleha, Muhammad Sharif, Afifa Masood, Mussarat Yasmin, and Mudassar Raza. "A survey on medical image segmentation." *Current Medical Imaging.* 2015. 11(1):3–14.

Milosavljevic, Tomica, Mirjana Kostić-Milosavljević, Ivan Jovanović, and Miodrag Krstić. "Complications of peptic ulcer disease." *Digestive Diseases.* 2011. 29(5):491–493.

Mukherjee, Rashmi, Dhiraj Dhane Manohar, Dev Kumar Das, Arun Achar, Analava Mitra, and Chandan Chakraborty. "Automated tissue classification framework for reproducible chronic wound assessment." *BioMed Research International.* 2014:1–10.

Muñoz, P L, R Rodríguez, and N Montalvo. "Automatic segmentation of diabetic foot ulcer from mask region-based convolutional neural networks." *Journal of Biomedical Research and Clinical Investigation.* 2020. 1(1):1–9.

Mwawado, R H, B J Maiseli, and Mussa A Dida. "Robust edge detection method for segmentation of diabetic foot ulcer images." *Engineering, Technology and Applied Science Research.* 2020. 10(4):6034–6040.

Nanni, Loris, Stefano Ghidoni, and Sheryl Brahnam. "Handcrafted vs. non-handcrafted features for computer vision classification." *Pattern Recognition.* 2017. 17:158–172.

Naz, Javeria, Muhammad Sharif, Mussarat Yasmin, Mudassar Raza, and Muhammad A Khan. "Detection and classification of gastrointestinal diseases using machine learning." *Current Medical Imaging.* 2021. 17(4):479–490.

Nelson, E Andrea, and Matthew D Bradley. "Dressings and topical agents for arterial leg ulcers." *Cochrane Database of Systematic Reviews.* 2007. 1–22.

Nisa, Maryam, Jamal Hussain Shah, Shansa Kanwal, Mudassar Raza, Muhammad Attique Khan, Robertas Damaševičius, and Tomas Blažauskas. "Hybrid malware classification method using segmentation-based fractal texture analysis and deep convolution neural network features." *Applied Sciences.* 2020. 10(14):1–23.

Noor, Saba, Mohammad Zubair, and Jamal Ahmad. "Diabetic foot ulcer—a review on pathophysiology, classification and microbial etiology." *Diabetes and Metabolic Syndrome: Clinical Research and Reviews.* 2015. 9(3):192–199.

Ohura, Norihiko, Ryota Mitsuno, Masanobu Sakisaka, Yuta Terabe, Yuki Morishige, Atsushi Uchiyama, Takumi Okoshi, Iizaka Shinji, and Akihiko Takushima. "Convolutional neural networks for wound detection: The role of artificial intelligence in wound care." *Journal of Wound Care.* 2019. 28(10):13–24.

Pachouri, Kapil Kumar. "A comparative analysis and survey of various feature extraction techniques." *International Journal of Computer Science and Information Technologies.* 2015. 6(1):377–379.

Patel, Samsunnisha, Rachna Patel, and Dhara Desai. "Diabetic foot ulcer wound tissue detection and classification." *International Conference on Innovations in Information, Embedded and Communication Systems.* 2017. 1–5.

Paton, Joanne, Graham Bruce, Ray Jones, and Elizabeth Stenhouse. "Effectiveness of insoles used for the prevention of ulceration in the neuropathic diabetic foot: a systematic review." *Journal of Diabetes and its Complications.* 2011. 25(1):52–62.

Pounder, R E. "Model of medical treatment for duodenal ulcer." *The Lancet.* 1981. 317(8210):29–30.

Prompers, L, M Huijberts, Jan Apelqvist, E Jude, A Piaggesi, K Bakker, M Edmonds et al. "High prevalence of ischaemia, infection and serious comorbidity in patients with diabetic foot disease in Europe. Baseline results from the Eurodiale study." *Diabetologia.* 2007. 50(1):18–25.

Qureshi, Imran, Muhammad Sharif, Mussarat Yasmin, Mudassar Raza, and Muhammad Y Javed. "Computer aided systems for diabetic retinopathy detection using digital fundus images: A survey." *Current Medical Imaging.* 2016. 12(4):234–241.

Raza, Mudassar, Zonghai Chen, Saeed-Ur Rehman, Peng Wang, and Peng Bao. "Appearance based pedestrians' head pose and body orientation estimation using deep learning." *Neurocomputing.* 2018. 272:647–659.

Rodríguez-Esparza, Erick, Laura A Zanella-Calzada, Diego Oliva, Ali Asghar Heidari, Daniel Zaldivar, Marco Pérez-Cisneros, and Loke Kok Foong. "An efficient Harris Hawks-inspired image segmentation method." *Expert Systems with Applications.* 2020. 155:1–29.

Ronneberger, Olaf, Philipp Fischer, and Thomas Brox. "U-net: Convolutional networks for biomedical image segmentation." International Conference on Medical image computing and computer-assisted intervention, Springer, Cham. 2015. 234–241.

Tarnawski, A S., and A Ahluwalia. "Molecular mechanisms of epithelial regeneration and neovascularization during healing of gastric and esophageal ulcers." *Current Medicinal Chemistry.* 2012. 19(1):16–27.

Saba, Tanzila, Ahmed Sameh Mohamed, Mohammad El-Affendi, Javeria Amin, and Muhammad Sharif. "Brain tumor detection using fusion of hand crafted and deep learning features." *Cognitive Systems Research.* 2020. 59:221–230.

Saba, Tanzila, Ahmed Sameh, Fatima Khan, Shafqat Ali Shad, and Muhammad Sharif. "Lung nodule detection based on ensemble of hand crafted and deep features." *Journal of Medical Systems.* 2019. 43(12):1–12.

Saba, Tanzila, Amjad Rehman, Nor Shahida Mohd Jamail, Souad Larabi Marie-Sainte, Mudassar Raza, and Muhammad Sharif. "Categorizing the students' activities for automated exam proctoring using proposed deep L2-GraftNet CNN network and ASO based feature selection approach." *IEEE Access.* 2021. 9:47639–47656.

Saba, Tanzila, Amjad Rehman, Rabia Latif, Suliman Mohamed Fati, Mudassar Raza, and Muhammad Sharif. "Suspicious activity recognition using proposed deep L4-branched-ActionNet with entropy coded ant colony system optimization." *IEEE Access.* 2021. 9:89181–89197.

Santilli, Jamie D, and Steven M Santilli. "Chronic critical limb ischemia: diagnosis, treatment and prognosis." *American Family Physician.* 1999. 59(7):1899–1908.

Sathya, Bharath, and R Manavalan. "Image segmentation by clustering methods: performance analysis." *International Journal of Computer Applications.* 2011. 29(11):27–32.

Schaper, N C. "Diabetic foot ulcer classification system for research purposes: a progress report on criteria for including patients in research studies." *Diabetes/Metabolism Research and Reviews.* 2004. 20(1):590–595.

Sharif, Muhammad Imran, Muhammad Attique Khan, Musaed Alhussein, Khursheed Aurangzeb, and Mudassar Raza. "A decision support system for multimodal brain tumor classification using deep learning." *Complex and Intelligent Systems.* 2021. 1–14.

Sharif, Muhammad, Sajjad Mohsin, Muhammad Jawad Jamal, and Mudassar Raza. "Illumination normalization preprocessing for face recognition." The 2nd Conference on Environmental Science and Information Application Technology. 2010. 2. 44–47.

Sivayamini, L, C Venkatesh, S Fahimuddin, N Thanusha, S Shaheer, and P Sujana Sree. "A novel optimization for detection of foot ulcers on infrared images." International Conference on Recent Trends in Electrical, Electronics and Computing Technologies. 2017. 41–43.

Solís-Sánchez, Luis Octavio, J M Ortiz-Rodriguez, Rodrigo Castañeda-Miranda, M R Martinez-Blanco, G Ornelas-Vargas, Jorge I Galván-Tejada, Carlos Eric Galvan-Tejada, José M Celaya-Padilla, and Celina Lizeth Castañeda-Miranda. "Identification and evaluation on diabetic foot injury by computer vision." IEEE International Conference on Industrial Technology. 2016. 758–762.

Song, Bo, and Ahmet Sacan. "Automated wound identification system based on image segmentation and artificial neural networks." IEEE International Conference on Bioinformatics and Biomedicine. 2012. 1–4.

Souaidi, Meryem, and Mohamed El Ansari. "Multi-scale analysis of ulcer disease detection from WCE images." *IET Image Processing.* 2019. 13(12):2233–2244.

Sudarvizhi, D, M Nivetha, P Priyadharshini, and J R Swetha. "Identification and analysis of foot ulceration using load cell technique." *International Research Journal of Engineering and Technology*. 2019. 6:7792–7797.

Taghanaki, Saeid Asgari, Kumar Abhishek, Joseph Paul Cohen, Julien Cohen-Adad, and Ghassan Hamarneh. "Deep semantic segmentation of natural and medical images: A review." *Artificial Intelligence Review*.2021. 54(1):137–178.

Tang, Jun. "A color image segmentation algorithm based on region growing." 2nd International Conference on Computer Engineering and Technology. 2010. 6:1–34.

Treece, K A, R M Macfarlane, N Pound, F L Game, and W J Jeffcoate. "Validation of a system of foot ulcer classification in diabetes mellitus." *Diabetic Medicine*. 2004. 21(9):987–991.

Tremeau, Alain, and Nathalie Borel. "A region growing and merging algorithm to color segmentation." *Pattern Recognition*. 1997. 30(7):1191–1203.

Vali, Shaik Bajid, Anil Kumar Sharma, and Syed Musthak Ahmed. "Implementation of modified chan vase algorithm to detect and analyze diabetic foot ulcers." International Conference on Recent Trends in Electrical, Electronics and Computing Technologies. 2017. 36–40.

Varacallo, M, and S D Mair. "StatPearls [Internet] StatPearls Publishing." Treasure Island (FL). 2020. 1–20.

Vardasca, R, C Magalhaes, A Seixas, R Carvalho, and J Mendes. "Diabetic foot monitoring using dynamic thermography and AI classifiers." Proceedings of the 3rd Quantitative InfraRed Thermography Asia Conference, Tokyo, Japan. 2019. 1–5.

Veredas, Francisco, Héctor Mesa, and Laura Morente. "Binary tissue classification on wound images with neural networks and bayesian classifiers." *IEEE Transactions on Medical Imaging*. 2009. 29(2):410–427.

Voulodimos, Athanasios, Nikolaos Doulamis, Anastasios Doulamis, and Eftychios Protopapadakis. "Deep learning for computer vision: A brief review." *Computational Intelligence and Neuroscience*. 2018. 1–14.

Wang, Changhan, Xinchen Yan, Max Smith, Kanika Kochhar, Marcie Rubin, Stephen M. Warren, James Wrobel, and Honglak Lee. "A unified framework for automatic wound segmentation and analysis with deep convolutional neural networks." 37th Annual International Conference of the IEEE Engineering in Medicine and biology Society. 2015. 2415–2418.

Wang, Lei, Peder C Pedersen, Diane M Strong, Bengisu Tulu, Emmanuel Agu, and Ronald Ignotz. "Smartphone-based wound assessment system for patients with diabetes." *IEEE Transactions on Biomedical Engineering*. 2014. 62(2):477–488.

Wang, Lei, Peder C Pedersen, Emmanuel Agu, Diane M Strong, and Bengisu Tulu. "Area determination of diabetic foot ulcer images using a cascaded two-stage SVM-based classification." *IEEE Transactions on Biomedical Engineering*. 2016. 64(9):2098–2109.

Wannous, Hazem, Yves Lucas, and Sylvie Treuillet. "Enhanced assessment of the wound-healing process by accurate multiview tissue classification." *IEEE Transactions on Medical Imaging*. 2010. 30(2):315–326.

Wannous, Hazem, Yves Lucas, Sylvie Treuillet, and Benjamin Albouy. "A complete 3D wound assessment tool for accurate tissue classification and measurement." 15th IEEE International Conference on Image Processing. 2008. 2928–2931.

Wijesinghe, Isuru, Chathurika Gamage, Indika Perera, and Charith Chitrraranjan. "A smart tele-medicine system with deep learning to manage diabetic retinopathy and foot ulcers." Moratuwa Engineering Research Conference. 2019. 686–691.

Wild, Sarah, Gojka Roglic, Anders Green, Richard Sicree, and Hilary King. "Global prevalence of diabetes: Estimates for the year 2000 and projections for 2030." *Diabetes Care*. 2004. 27(5):10471053.

Williams, D T, Joanna Ruth Hilton, and Keith Gordon Harding. "Diagnosing foot infection in diabetes." *Clinical Infectious Diseases*. 2004. 39(2):583–586.

World Health Organization. *World health statistics: monitoring health for the SDGs sustainable development goals*. World Health Organization. 2016. 1–121.

Wu, Cheng-Shyong, Chun-Yen Lin, and Yun-Fan Liaw. "*Helicobacter pylori* in cirrhotic patients with peptic ulcer disease: a prospective, case controlled study." *Gastrointestinal Endoscopy*. 1995. 42(5):424–427.

Wu, Jian Kang, Mohan S Kankanhalli, Joo-Hwee Lim, and Dezhong Hong. "Color feature extraction." *Perspectives on Content-Based Multimedia Systems*. 2000: 49–67.

Yap, Moi Hoon, Choon-Ching Ng, Katie Chatwin, Caroline A Abbott, Frank L Bowling, Andrew J M Boulton, and Neil D Reeves. "Computer vision algorithms in the detection of diabetic foot ulceration: A new paradigm for diabetic foot care?" *Journal of Diabetes Science and Technology*. 2016. 10(2):612–613.

Yap, Moi Hoon, Katie E Chatwin, Choon-Ching Ng, Caroline A Abbott, Frank L Bowling, Satyan Rajbhandari, Andrew J M Boulton, and Neil D Reeves. "A new mobile application for standardizing diabetic foot images." *Journal of Diabetes Science and Technology*. 2018. 12(1):169–173.

Yu, Hongshan, Zhengeng Yang, Lei Tan, Yaonan Wang, Wei Sun, Mingui Sun, and Yandong Tang. "Methods and datasets on semantic segmentation: A review." *Neurocomputing*. 2018. 304:82–103.

Yusuf, Nurlisa, Ammar Zakaria, Mohammad Iqbal Omar, Ali Yeon Md Shakaff, Maz Jamilah Masnan, Latifah Munirah Kamarudin, Norasmadi Abdul Rahim et al. "In-vitro diagnosis of single and poly microbial species targeted for diabetic foot infection using e-nose technology." *BMC Bioinformatics*. 2015. 16(1):1–12.

8 Review of Deep Learning Techniques for Prognosis and Monitoring of Diabetes Mellitus

C. Muthamizhchelvan
SRM Institute of Science and Technology, Tamil Nadu, India

K.A. Sunitha
SRM Institute of Science and Technology, Tamil Nadu, India

M. Saranya
Department of Electronics and Instrumentation
Engineering, SRM Institute of Science and Technology,
Tamil Nadu, India

B. Venkatraman
Resource Management and Public Awareness Group,
Indira Gandhi Centre for Atomic Research, Kalpakkam, India

M. Menaka
Health Safety and Environment, Indira Gandhi Centre for
Atomic Research, Kalpakkam, India

Sridhar P. Arjunan
SRM Institute of Science and Technology, Tamil Nadu, India

8.1 INTRODUCTION

8.1.1 PREVALENCE OF DIABETES MELLITUS

Diabetes mellitus (referred to as diabetes) is a category of diseases that can cause significant human and economic damage [1]. It has become an increasing global burden, with 451 million people diagnosed in 2019 and 693 million expected to be diagnosed by 2045. There are less than half a billion diabetic people in the world and in the future, the expected rise will be 25% in 2030 and 51% in 2045, as depicted in

DOI: 10.1201/9781003230540-8

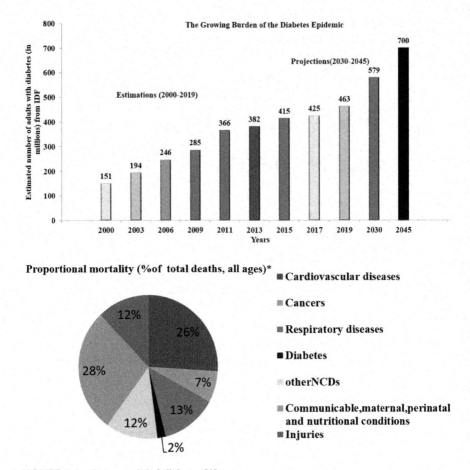

FIGURE 8.1   Data on global diabetes [1].

Figure 8.1 [2]. Diabetes is recognized as one of the largest worldwide health emergencies of the century and is among the top ten causes of death. According to the World Health Organization (WHO), one person in the world is projected to die every second due to diabetes or its complications, with half of those deaths (4 million/year) happening in those under the age of 60. In India, diabetes contributes to 2% of deaths.

8.1.2   REVIEW OF DIABETES MELLITUS

Diabetes is an incurable autoimmune disease caused by a lack of or resistance to the hormone insulin [3]. The pancreas produces an important hormone that enables cells of the body to consume glucose in the blood. Blood sugar is derived from the digested food and is essential to provide the energy necessary for the body to function [4]. When there is a lack of insulin or resistance to insulin, body cells are unable to absorb blood sugar and the level of sugar increases in the blood. The presence of high blood

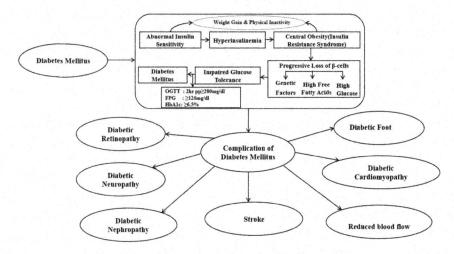

FIGURE 8.2 Summary of diabetes diagnosis and monitoring.

sugar levels in the blood is known as hyperglycemia. When uncontrolled, high blood sugar levels lead to various complications of diabetes, as shown in Figure 8.2.

8.1.3 Types of Diabetes

Diabetes is classified into three clinical classes:

Type 1 diabetes is due to destruction of the pancreatic beta-cell and results in total insulin deficiency [5].

Type 2 diabetes occurs as a result of insulin secretory deficiency in the presence of insulin resistance. Some types of diabetes are caused by other conditions, such as genetic deficiencies in cell function and insulin action, exocrine pancreas issues, or medication or chemical-mediated diabetes.

Gestational diabetes is diagnosed during pregnancy [6].

8.1.4 Diabetes Mellitus and COVID-19 pandemic

Diabetes is a common co-morbidity in COVID-19 patients. Following COVID-19, several complications have been reported for diabetes patients with elevated levels of pancreatic damage from India and other countries. Some of the cases included the development of diabetic ketoacidosis and type 1 diabetes symptoms. Angiotensin-converting enzyme-2 (ACE-2), a receptor found in pancreatic beta-cells, allows SARS-CoV-2 to enter the cell, cause cell damage, and ultimately lead to the development of diabetes [7]. Patients with newly diagnosed diabetes with a history of COVID-19 infection (self-reported history) were reported to have higher fasting blood glucose, postprandial blood glucose, and insulin levels. The COVID-19 infection that presented with diabetic ketoacidosis resolved after initial insulin therapy and glycemia was controlled with oral anti-diabetes drugs [8]. As a result, during this pandemic period, there is a risk of an increase in various lifestyle-related diseases such as new-onset diabetes. The effects of lockdown have worsened the lifestyle of people, as evidenced in each individual progressing from pre-diabetes to diabetes [9].

More intricate research and randomized control preliminaries are needed to explain unclear problems concerning the molecular and therapeutic interrelationship between diabetes and COVID-19 infection [10]. In countries where the burden of diabetes is already high, many people suffering from diabetes became more vulnerable and sensitive to COVID-19 infection. As most of these patients already had various complications related to diabetes, they were at increased risk of COVID-19 infection and the prognosis was bleak due to several complications. In any case, prevention and early detection of diabetes are essential to ensure that the burden due to the disease can be alleviated.

8.1.5 CLINICAL AND RESEARCH TOOLS FOR DIABETES

Diabetes has been diagnosed for decades based on blood glucose levels. Several researchers have examined the history of glucose detection system design, and a list of four decades of glucose monitoring [9], classified by technology used, is shown below. In the 1970s, the first-generation glucose monitors that used reflectance technology were tested, and those instruments permitted a comparatively larger amount of blood for measurement. The second-generation devices used a blood drop and were designed as simpler, more portable devices for personalized use due to available technologies. The sensors of the third generation were launched as minimally invasive devices with a range of tiny skin needles, allowing continuous glucose monitoring (CGM) [9]. Fourth-generation medical devices provide non-invasive methods that allow remote and real-time continuous monitoring.

Table 8.1 shows the results of various research studies on invasive and non-invasive techniques for blood glucose monitoring. Irrespective of whether the technique is invasive or non-invasive, the accuracy of detecting diabetes or monitoring blood glucose level is the primary outcome.

These research outcomes confirm that the identified biomarkers are prominent biomarkers to be used for the early diagnosis of diabetes. Hence, a deep learning (DL) algorithm can use data from these prominent biomarkers and help in both disease diagnosis and prognosis.

8.1.6 VITAL ROLE OF DEEP LEARNING IN DIABETES MELLITUS

DL is superior to conventional machine learning because the algorithm can utilize raw data and learn from the data using several layers of computations. This allows the algorithm to learn abstractions based on inputs [13]. The need for this research is to determine or develop a DL procedure that will aid in an early analysis of the pre-diabetes stage that will postpone the onset of diabetes and thereby its complications. A review of DL-based analytic methods that can assist clinicians in the prognosis and diagnosis of diabetes with various biomarkers is discussed in this chapter. Further, many invasive and non-invasive methods for early diagnosis and monitoring of diabetes, which can be implemented using a DL technique for both clinical and research

TABLE 8.1
Clinician assessment tools for diagnosis of diabetes mellitus

Clinical tool/ invasive technique	Techniques	Advantages	Disadvantages	Parameters/results/remarks		
				Sensitivity	Specificity	Accuracy
	Blood sugar meters	Cost-effective [3]	Precision and unreliability	Sensitivity is 81%	Specificity is 65%	Accuracy 74%
	Blood lancets	Less pain. More practical to use in children [3]	Significant in diagnosis	Sensitivity is high	Specificity is high	Decreases
	Diabetic test strips	The concentration in blood and glucose are similar [3]	Long and complex calibration methodology, sweat and temperature sensitivities, large size	Sensitivity is low	Specificity is low	Accuracy is moderate level
	Ketone test strips	Rapid measurement [11]	Delay in display results	Sensitivity and superior specificity to urine ketone testing, dilution methods if necessary to ensure accuracy, and technologies need to be developed to minimize the number of substances that interfere with the data		
	Blood glucose levels	Simple utilization [12]	Precision and accuracy of the devices are unreliable	Sensitivity 96%	Specificity 66%	Accuracy 82%

tools, have been included. Specifically, various DL research strategies for prognosis and diagnosis of diabetes using biomarkers, such as retinal fraction dimensions, skin temperature, and exhaled volatile organic compounds, are discussed (Table 8.2). So, this chapter discusses in depth the use of DL methods in the diagnosis and prognosis of diabetes.

8.2 UNDERSTANDING OF DEEP LEARNING-BASED DIAGNOSTIC SYSTEMS APPLICABLE TO DIABETES MELLITUS

DL is emerging as one of the primary methods for predictive diagnosis in diseases such as diabetes. Deep-view DL models and algorithms assist in DM studies in diagnosis, treatment, diabetic retinopathy (DR), and biomarker recognition.

A wide range of therapeutic techniques is inextricably tied to the exponential advancement of DL in medicine. Seeking the right way to extend DL to all levels of medical care becomes more complicated. It is based on two factors: continued advancement in technologies and accumulation of medical knowledge [17].

Artificial intelligence (AI) is a scientific prediction system that systematizes learning and recognizes patterns for diagnosing pre-diabetes and diabetes [18]. A few blood tests can be done to detect diabetes, including:

1. Fasting plasma glucose: this assists in predicting blood glucose levels after an 8-hour fast.
2. HbA1C test: this assists in predicting glucose levels over a quarter-year period.
3. Blood tests, as well as a glucose challenge test and a 3-hour glucose resilience test, are performed in the 24th and 28th weeks of pregnancy to detect gestational diabetes [19].

However, these experimental procedures usually take a long time to achieve results which may be inconvenient for patients. Given the difficulties of diagnosing diabetes using traditional medical methods, the development of a non-invasive and convenient automated healthcare system to diagnose and avoid these diseases is both important and desirable.

Several experiments have recently focused on forecasting diabetes using machine learning and DL methods. For example, using the deep neural network (DNN) approach, researchers recommended a DL-based technique for diabetes data classification. The suggested scheme was tested using the Pima Indian diabetes data collection. The proposed scheme demonstrated good recognition accuracy (86.26%), demonstrating DNN's efficacy in assisting doctors in disease prediction. As shown in Figure 8.3, DL methods such as convolutional neural network (CNN), long short-term memory networks, recurrent neural network (RNN), stacked auto-encoder, deep belief networks, deep Boltzmann computer, and others have been used for diagnosis, tracking, classification, and risk evaluation of diabetes for over 20 years. DL, on the other hand, incorporates high-level imaging highlights from large amounts of an image that are appropriate for training purposes [20].

TABLE 8.2
Research assessment tools for diagnosis of diabetes mellitus

Research tool/ non-invasive technique	Techniques	Advantages	Disadvantages	Parameters/results/remarks					
				Sensitivity		Specificity		Accuracy	
	HbA1c – Glycated form of hemoglobin (HbA1c is a test to measure the amount of glycated hemoglobin in blood)	Better results, minimize healthcare cost	Predicting the risk of future type 2 diabetes	Sensitivity	78.6%	Specificity	72.5%	Accuracy	More precise
	Fractal Dimension (FD) measurement tools – FD is a statistical index of complexity that compares how the detail in a pattern changes as the scale at which it is measured changes	Each pixel in the image is assigned an absolute curvature value using this method [14]	Joint segmentation results in the difference in diabetic retinopathy grade results	Sensitivity	93.7%	Specificity	92.9%	Accuracy	94.8%
	Thermal imaging camera tools – Thermal radiation can be used to detect heat in the human body	Safe decision based on results [15]	High-risk indicator for abnormal infrared pattern images	Sensitivity is high		Specificity is low		Accuracy is high	

(continued)

TABLE 8.2 Cont.
Research assessment tools for diagnosis of diabetes mellitus

			Sensitivity is high	Specificity is available technology limits the ability to achieve the required specificity/selectivity	Accuracy suffers from a lack of specificity and inaccuracy as a result of subject movement and sweating, skin irritation, and so on
Breath biomarkers – Breath biomarkers detect early diagnosis progress, success rate, and improve the efficiency of drug development	Non-invasive: repeated quantities are permitted. Monitoring in real time is possible [15]	In multi-component mixtures, it is not very useful for identifying unknown compounds			
Fundus camera (FD) – A low-power microscope with a camera attached. FD is described by the angle of view [16]	Clear images are acquired	Cost and skilled professionals required to operate	Sensitivity	Specificity	Accuracy
			61–90%	85–97%	96.5%

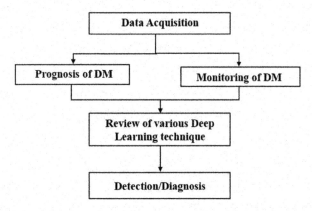

FIGURES 8.3   The overall flow of the chapter discussion of diabetes in medical diagnosis using deep learning.

8.3   LITERATURE REVIEW: IMPLEMENTATION OF DEEP LEARNING ALGORITHM FOR THE PROGNOSIS OF DIABETES MELLITUS USING EARLY BIOMARKERS

A prognostic is research that aids in the prediction of an occurrence before it happens, allowing for more effective crucial decision making. Front-line doctors may use prognostics to predict how an illness will affect a patient and respond appropriately to save the patient's life. Machine learning techniques may produce more precise outcomes, enabling one to understand better the effects of these functions in terms of diabetes development. An effective clinical solution based on these characteristics will prevent the development of diabetes and thereby increase the survival of patients [9].

This section discusses numerous early biomarkers that were developed using DL algorithms for the early detection of diabetes [9]. Multiple diabetes parameters from image processing cohering with the development of diabetes as an early biomarker are still under review.

A new study has identified diabetes biomarkers such as blurred visualization, retinal vasculature [9], and skin temperature [21] that can be used for early detection. The significance of studying volatile organic compounds (VOCs) released by human exhaled breath [22] and wound ulcer healing status [23] as an early biomarker for diabetes is currently unexplained.

8.3.1   BIOMARKERS FOR EARLY DIAGNOSIS: RETINAL FRACTAL DIMENSIONS

Using the fractal dimension (FD) approach, this technique explores and computes deviations in the neural forms of the retina vascular network in diabetes. DR is currently the primary reason for blindness in middle-aged individuals. It is predicted that over 93 million people [23] will be affected. If unchecked, DR, an eye disease associated with long-term diabetes, can lead to blindness. If diagnosed early, DR may be treated or reversed with treatment or laser procedures. Diabetic retinopathy signs do not appear until permanent damage has occurred, so early diagnosis is important.

If DR is diagnosed early, the progression of vision deficiency can be delayed or avoided; however, this can be problematic since the condition sometimes presents few signs until it is too late to have adequate care. Clinicians may recognize DR by the appearance of lesions associated with the disease's vascular anomalies. While this solution is useful, it has a high capital need. The requisite skills and equipment are often unavailable in areas with a high prevalence of diabetes in local communities and where DR diagnosis is most needed. If the number of diabetic patients increases [24], the facilities required to avoid DR-related blindness will become much more inadequate. The need for a full and automatic DR screening process is well known and previous attempts have been made to achieve good results utilizing image classification, pattern recognition, computer education, and DL, with color fundus photographing as an input. To optimize the effect a model can have on improved DR identification, the best models are possible.

Retinal IFDs, a measure of the vasculature patterns, namely tortuosity, and the vessel caliber summary help us to document tiny changes in the retinal vasculature at the earliest opportunity, using a fundus camera attached to vasculature assessment software, thereby acting as an early biomarker in DR [24]. Rich features from retinal fundus images can be extracted using the DL technique.

8.3.1.1   Retinal Vascular Biomarkers for Early Detection and Screening

Changes in the morphological and topological structure of the vessels could represent changes in blood flow and vessel blood pressure (Figure 8.4) in four categories. The emphasis is on the construction of a detailed, quantitative study of retinal microwaves: vessel diameter, tortuosity of vessels, vessel bifurcations, and vessel fractal measurements. Many vascular biomarkers are used to diagnose even the smallest early improvements in the quantitative study of the bladder: the diameter of the vessel, vessel tortuosity, bifurcation-based characteristics, and FD [20].

8.3.1.2   Fractal Dimensions

Benoit Mandelbrot (French-American mathematician) initially introduced the term "fractal" in his 1975 book *Les Objets Fractals*. Fractal is appropriate for non-smooth or uneven sets. The general characteristics of a fractal are identity, affinity, reversal, and quadrature [25], as shown in Table 8.3.

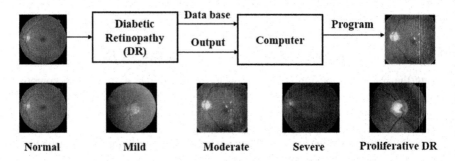

FIGURE 8.4   Image performance of the retinal fundus [45].

TABLE 8.3
The characteristics of fractals

Self-similarity	Similar different scale
Self-affinity	Parts integrated
Self-squaring	The inverse of the previous fractal shapes at the complex level
Self-inversion	Inverted version of the other arrangement

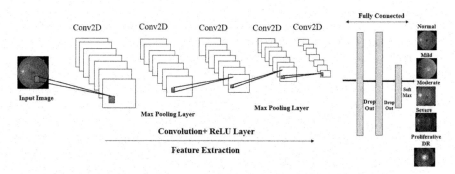

FIGURE 8.5   Deep learning using diabetic retinopathy.

In literature, there are potential methods for calculating the retinal FD of fundus images using various methods, such as box counting, Hausdorff FD (HFD), modified fractal, and Fourier fractal. A study that estimated the coefficient of variation for all FD methods was developed for the analysis of retinal images; it determined that the HFD method had the least coefficient of variation [24]. The two different types of FD analysis method in box counting were investigated: the modest and standard method to evaluate the FD of fractal objects calculating the FD dimension in diabetes in retinal image value is 1.446 and that in non-diabetic retinal image is 1.502. Direct implementation of the Hausdorff dimension in diabetic retinal images is 1.541 and non-diabetic retinal image value is 1.544. The outcome of fractal analysis for the diabetes/non-diabetes retinal image suggests that classic FDs must be calculated under strict conditions, and tiny changes in the images and vessel segmentation can cause significant variations. The comparison of FD in diabetic retinal image value is less than the non-diabetic retinal image value. The results show that the general errors in FD estimation are critical, and FD varies dramatically depending on image quality, methodology, and the strategy used to estimate it. The estimation of FD using automated and semi-automated strategies is not stable enough, making FD a misleading biomarker in quantitative clinical applications [21], [25].

Using DL techniques, we can detect glaucoma earlier and stop the progression of macular degeneration, diabetic retinopathy, and various eye-related issues before the loss of vision. Both quantitatively and qualitatively, the Hausdsoff FD is the "best" approach for computing FD for retinal vasculature (Figure 8.5). The therapeutic importance of the retinal FD is evaluated using current science data on the interaction between FD [25] and retinal pathology.

8.3.2 SKIN TEMPERATURE

Foot ulcers are a major complication of diabetes and are the main reason for diabetes amputation. Patients with diabetes have an approximately 12.25% lifetime chance of developing foot ulcer for primarily in connection with peripheral neurology and also peripheral arterial disorder. Peripheral neuropathy is a frequent complication of diabetes that particularly affects the nerve extremities. Peripheral neuropathy of the lower extremity affects approximately 85% of diabetic patients [46]. Differing nervous system divisions such as the motor, sensory, and autonomous can cause nerve damage.

8.3.2.1 Diagnosis with Medical Infrared Thermography

Human skin can be called a "blackbody radiator" because it physiologically emits infrared radiation. A thermal imagery sensor can detect this infrared radiation. Thermal infrared imagery operates by random measurements of infrared radiations released by different parts of the body. The thermal infrared pattern is dependent on the microcirculation on the surface of the skin. Thermal imagery can also be used to detect time-related changes to microcirculation. These features make the infrared thermal imagery a perfect candidate for early detection of pathological conditions that change the microcirculation and thereby emission of infrared radiation. Pathological conditions include foot ulcers and neuropathy due to diabetes, cancer of breast and lung, arthritis, sports injury, glaucoma, tooth abscess, lesions in the skin, respiratory issues, and Raynaud's phenomenon [26].

As shown in Figure 8.6, there are several effective computational approaches in the diagnosis and treatment of medical conditions. This figure shows a standard diagnostic device that employs infrared thermography, image processing, and computing to detect a disease.

The distinctive characteristic of thermography is to picture the appearance of the skin vasculature rather than the anatomy of the structure, unlike other imaging methodologies. The application of thermal image registration and analysis in medicine has also been mentioned. A study on algorithms used in computer-aided diagnostic systems using thermography has been conducted [15]. Techniques for denoising infrared pictures, extracting a backdrop from infrared images to distinguish areas of interest, and statistical analysis of regions of interest to differentiate between regular and irregular temperatures, are currently available.

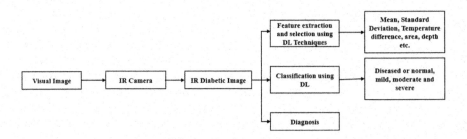

FIGURE 8.6   Use of soft computing and infrared thermography for disease diagnosis.

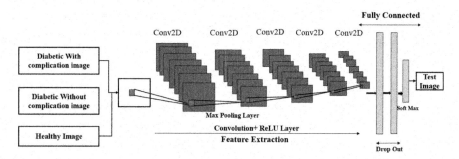

FIGURE 8.7 Convolutional neural network architecture for the prognosis of skin thermal images.

8.3.2.2 Role of Deep Learning in Infrared

Current diagnostic systems on thermal infrared image processing are based on various soft computing techniques such as neurocomputing, vague logic, probabilistic computing, evolutionary computing, and machine learning components. Some of the methods incorporating these soft computation skills include artificial neural networks (ANN), fuzzy-C-means (FCM), and Bayesian networks. According to Safitri et al. [47], DL structural designs are leading the way for more precise autonomous systems, due to advances in hardware and computational capability such as graphics processing units (GPUs). For medical thermal infrared image processing, several hybrid DL approaches have been used (Figure 8.7).

In biomedicine, infrared thermography is the most promising technique among other conventional methods for revealing differences in skin temperature caused by irregular temperature dispersion; such temperature dispersions signal the presence of diseases and disorders [27].

8.3.3 Exhaled Volatile Organic Compounds

Exhaled breath analysis and processing have long kindled the interest of those working in medical diagnosis and illness monitoring. The advantage of the method is its non-invasive design, ample availability (i.e., breath), and ability to facilitate prompt at-patient identification. Gases are exhaled as byproducts of human cell metabolism or that of microorganisms from the pulmonary tract, blood, or peripheral tissues.

VOCs are appealing candidates for biomarkers regardless of origin or physiological function. Exhaled VOCs are appealing candidates as biomarkers of cell process or metabolism, regardless of origin or physiological function, and may be integrated into potential non-invasive clinical testing instruments [28].

Breath bioinformatics is premised on the idea that when a person transitions from a stable to a pathological state, their VOC profile changes, and this can be identified and theoretically used for detection and surveillance [29]. Exogenous VOCs comprise the major component of the exhaled breath [30], but more important than the exogenous VOC is that created endogenously and from microbial colonization in the

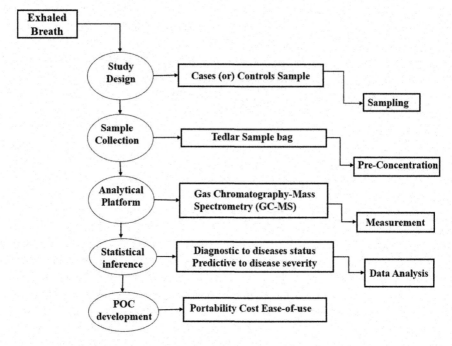

FIGURE 8.8 An off-line breath sampling pipeline.

body, as these are the clinically relevant VOCs. Endogenous VOCs help in identifying the physiological status of a patient while microbial VOCs help to determine different kinds of pathogens in the body and host–pathogen interaction. The main benefit of using breath VOCs is that they can be detected in very small amounts, such as parts per million volume (ppmv) or less, increasing detection precision. Figure 8.8 shows the different stages of this method:

- Sampling utilizes a bag for the gas sample, face mask, the Bio-VOCTM device, an apparatus to collect breath, and a canister.
- Preconcentration involves solid-phase micro extraction (SPME), thermal desorption (TD) tube, and needle trap devices (NTDs).
- Measurement includes gas chromatographs (GC) and mass analyzers [28].
- Data analysis includes data pre-treatment, processing, and analysis for both targeted and untargeted data.

The choice of biomarker plays an important role in the collection of samples and cost. The utility of late expiratory breath is that it has a high endogenous contribution [28]. These samples can be collected with certainty using available controls, such as a pre-determined late expiratory breath procedure. This level of diligence is required as each person's breath profile can differ due to several reasons [31].

8.4 LITERATURE REVIEW: PERFORMANCE OF DEEP LEARNING FOR DIAGNOSIS OF INVASIVE AND NON-INVASIVE METHODS FOR MONITORING DIABETES MELLITUS

There have been several methods of minimal to non-invasive monitoring of diabetes in the past decades. The high mortality and morbidity of diabetes have resulted in the development of different modalities of blood glucose monitoring [10].

8.4.1 BLOOD TEST

A glucometer is the current gold standard for self-monitoring of blood glucose levels. The device consists of a lancet, optical disposable strips, and a glucometer. A patient's finger is pricked with the lancet to derive a drop of blood that is placed on the optical disposable strip. This strip is inserted into a glucometer that determines and displays the blood glucose value. This helps determine if a patient is hyperglycemic or hypoglycemic and initiate appropriate treatment [9].

Continuous invasive glucose monitoring

CGM is the most recent blood glucose monitoring technology. Patients may acquire US Food and Drug Administration (FDA)-approved devices that consist of one-use disposable electrochemical sensing components that can be worn for a time ranging from 72 hours to 7 days, depending on the design [32],

Mhaskar et al. 2017) [32] investigated the application of a DNN for the diagnosis of diabetes using blood glucose monitoring levels with 30% and 50% of training data values. Greater accuracy of 94.39% and error percentage value of 3.57 were obtained for hypoglycemia when 50% of patients used flat low-pass filtering for training. Hypoglycemia blood glucose ≤ 70 mg/dl accuracy was 88.72% and percentage of error value was 6.79. In the second type of glycemia range, where blood glucose 70–180 mg/dl, accuracy was 80.32% and error was 2.32, and finally in the hyperglycemia value, where blood glucose >180 mg/dl, accuracy was 64.88% and percentage of error value was 13.22. These results suggest that the larger the dataset, the greater the accuracy and the lower the error.

Non-invasive glucose monitoring

The novelty of the non-invasive process started with the launch of the Arise CGM in 2007 by Glucon (Boulder, CO) to measure glucose levels; this technology employs the "photoacoustic properties of blood." A sensor is attached to a blood vessel. This sensor emits short laser waves of sound and pressure that are absorbed by the surrounding tissue. The resultant increase in body temperature creates "an acoustic strain impulse" on the surface of the tissue. This impulse can be used to study the properties of the skin's intrinsic structure.

8.4.2 THE SPECTRAL PROPERTIES OF WOUND HEALING

One of the most difficult complications of clinical medicine is delayed wound healing in diabetic patients since it increases the chance of gangrene, amputation, and even death [33]. Chronic lesions, like diabetic ulcers and pressure ulcers, are the main well-being and financial burden for both patients and the healthcare system. Visual assessment by an experienced clinician has been used to assess all cases of cutaneous wounds [34].

8.4.3 RETINOPATHY

DR is the primary reason for blindness in diabetic patients aged 20–74 years. At the time of diagnosis, up to 21% of people with type 2 diabetes have retinopathy [16], and the others experience a degree of retinopathy over time. Medical specialists usually use fundus or retinal photographs of the patient's eyes to assess the severity and degree of retinopathy involved with diabetes. Since the number of diabetic subjects is swiftly increasing, the quantity of DR datasets provided by screening programs will also rise, placing a huge labor-intensive workload on medical experts and increasing the expense of healthcare services. Previously it had been reported that the use of DL systems in automatic DR identification resulted in high sensitivity and specificity. M. T. Esfahani et al. used a well-known CNN, ResNet34 [16], in their study to classify DR images from the Kaggle dataset as regular or DR images.

Table 8.4 summarizes the literature on DL techniques with good sensitivity and specificity. Y. Liu et al. [48] created a weighted-path CNN (WP-CNN) to detect

TABLE 8.4
Images of wound-type model extraction using deep learning methods

Extracted attributes	Method(s) applied	Dataset	Limitations	References
Wound area and dimension	Convolutional neural networks that are fully convolutional (FCNs)	Own diabetic foot ulcer (DFU) dataset with 705 images	The network is unable to detect small wounds and prefers to draw smooth contours, making it difficult to perform accurate segmentation	[36]
Wound area	Vanilla FCN architecture	The NYU wound database contains over 8000 high-resolution images	The average dice accuracy is 64.2%	[35]
Wound area	Fully convolutional network in vanilla	950 images of wounds	Instead of learning from human specialists, the watershed algorithm labels wounds in images	[35]
Wound area	Combination of traditional Methods and deep neural network (Mobile Net)	950 images of wounds	Effects of level segmentation and a multifaceted model	[35]

referable DR videos. Over 60,000 photographs were classified as referable or non-referable DR and enhanced multiple times to ensure that the classes were balanced. The images were resized to 299 × 299 pixels and normalized before being fed into the algorithm. The WP-CNN of 105 layers significantly outperformed the pre-trained ResNet [16], SeNet [35], and Dense Net [34] algorithms, with 94.23% precision in their dataset and 90.84% in the STARE dataset.

8.4.4 DIABETIC CARDIOMYOPATHY

The American Heart Association, the American College of Cardiology Foundation, and the European Society of Cardiology in collaboration with the European Association for the Study of Diabetes identified diabetic cardiomyopathy in 2013, as shown in Table 8.5. According to clinical trials, the percentage of heart failure in patients with diabetes ranges from 19% to 26%. Diabetic patients have a higher rate of heart disease (39%) than non-diabetic patients (23%), with a relative chance of 1.3 of experiencing heart failure following 43 months of observation [17].

Table 8.6 shows techniques for detecting diabetes cardiomyopathy with good accuracy. On patients with no history of diabetes or cardiovascular disease, Xiao et al. used ten-fold cross-validation of machine learning algorithms [37]. DL algorithms such as Glmnet, RF, XGBoost, and LightGBM can be used for the clinical prediction of type 2 diabetes. In DL, the multilayer feedforward network algorithm is implemented for effective early determination of diabetes risk. Leon and Maddox proposed an improved DNN-based diabetes risk prediction model that not only predicts but also identifies who will develop the disease [17]. A mixed model that used an algorithm based on both genetic predisposition and extreme machine learning has been developed for the diagnosis of type 2 diabetes with an accuracy of 97.5%.

8.5 PRELIMINARY RESULTS FROM DATA ANALYSIS

As per International Clinical Diabetic Retinopathy Disease Severity Scale, the severity of DR was divided into five stages (non-DR, mild non-proliferative DR (NPDR), moderate NPDR, severe NPDR, or proliferative DR (PDR), respectively). Mild NPDR is distinguished by the presence of microaneurysms. Moderate NPDR is defined as something more than microaneurysms but less than extreme NPDR due to the presence of cotton-wool spots (CWS), hard exudates, and retinal hemorrhages, as shown in Figure 8.9.

According to the standard of early treatment for diabetic retinopathy, diabetic macular edema is identified if hard exudates are detected within 500 μm of the macular center. Generally, diabetic subjects are more likely to have more severe symptoms and complications when infected with any virus, as particularly is the case for those who are affected with the COVID-19 virus during this pandemic. Increasing blood sugar deteriorates immunity and makes it vulnerable to infection. The mutated variant of the virus (mucormycosis) almost invariably affects immunocompromised individuals, especially those with diabetes. Black fungus infections are known as mucormycosis, and this condition appears to predominantly affect diabetics who have recovered from COVID-19.

TABLE 8.5
Deep learning techniques for classification of cardiovascular diseases

Dataset (dataset size)	Deep learning method	Lesion detection	AUC	Sensitivity	Specificity	Accuracy	References
CNN: Inception v3	EyePACS-1 and Messidor-2 (1748)	No	–	96.1%	93.9%	–	[38]
	EyePACS-1 and Messidor-2 (1748)			97.5%	93.4%		
CNN	Messidor-2 (1748)	Yes	0.980	96.8%	87.0%	–	
CNN: ResNet34	Kaggle (35000)	No	–	86%	–	85%	[16]
CNN: Inception Net V3, and the Alex Net	Kaggle (166)	No	–	–	–	AlexNet: 37.4%, VGG16: 50.03%, and InceptionNet: 63.23%	[39]
CNN: VggNet, AlexNet, ResNet, and GoogleNet	Kaggle (35126)	No	The greater VggNet-s, the better (0.9786)	VggNet-16 outperformed the other models (90.78%)	The greater VggNet-s, the better (97.43%)	The higher is VggNet-s (95.68%)	[34]
Fully CNN	STARE (20), DRIVE (40), and CHASE DB1 are all available (28)	No	0.9905	0.8315	0.9858	0.9694	[40]
			0.9821	0.8039	0.9804	0.9576	
			0.9855	0.7779	0.9864	0.9653	
A large residual network	HEI-MED (169) and E-optha (82)	Exudates (EX) only	0.9647	0.9227	–	–	[21]

CNN (ResNet50)	Messidor (1200) and IDRiD (516)	No	96.3%	92%	–	92.6%	[35]
CNN (WP-CNN, ResNet, DenseNet, and SeNet) is a type of neural network	Their own dataset (60000) and STARE (131)	No	0.9823 0.951	90.94% –	95.74%	94.23% 90.84%	[34]
CNN: modified Alexnet	Messidor (1190)	No	–	92.35	97.45	96.35	[41]

TABLE 8.6
The deep learning methods used for diabetic retinopathy detection/classification [49]

Method	Classification	Accuracy	References
Markov blanket estimation	Backpropagation of statistical errors in a multilayer neural network	97.92%	[18]
Dynamic Bayesian network	A neural network with a sigmoid transfer function that uses backpropagation	97.61%	[42]
Naive Bayesian network	Scaled conjugate gradient and Levenberg-Marquardt learning algorithms are used to train feedforward neural networks with sigmoid transfer functions	79.56%	[43]
Naive Bayesian network	The sigmoid activation function is used in a backpropagation-trained feedforward multilayer perceptron	76.95%	[44]
Naive Bayesian network	A sigmoid transfer function is used for the multilayer perceptron neural network	74%–90%	[42]

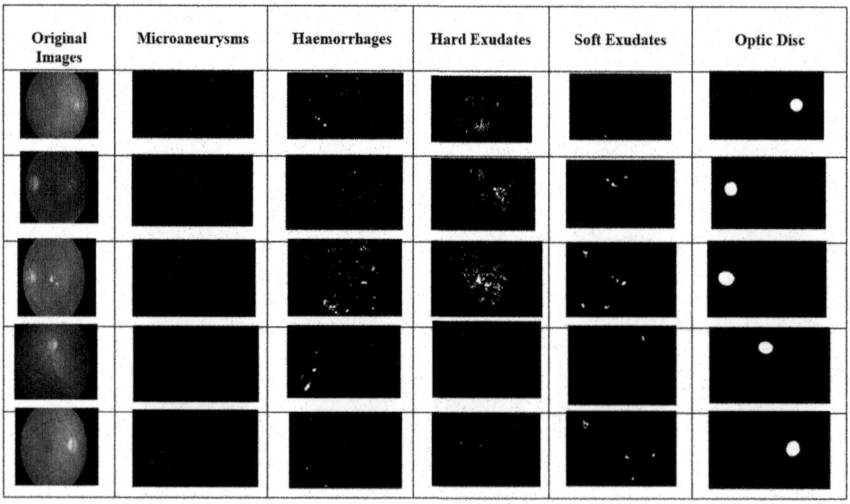

FIGURE 8.9 Retinal image feature extraction of deep learning algorithms [45].

Overall, 3,362 images from the Indian Diabetic Retinopathy Image Dataset (IDRiD), were used in this study. The dataset was divided into 80% for training and 20% for testing. Classes in the dataset were no diabetic retinopathy (NDR), mild, moderate, severe, and PDR. The main features representing these classes are microaneurysms, hemorrhage, soft and hard exudates, and optic disc. A DL technique using CNN architecture was used for the classification of images [45].

Severe and no DR categories had better performance in terms of accuracy, which was 95.74% and 94.75% respectively. A pathology-specified image possesses improved accuracy, sensitivity, specificity, false-negative, and false-positive rate. Thus, DL expending CNN architecture can determine the diagnostic differences between normal and abnormal retinal fundus images. The ground truth images provide boundary markings and give information about the eye. The dataset is intended to foster active research in DL.

Initially, CNN performs classification with better similarity measures using the sequence of deep neural layers. Output layer CNN architecture consists of binary cross-entropy categorical events to classify normal and abnormal retinal images with an accuracy of 98.5% in Figure 8.10.

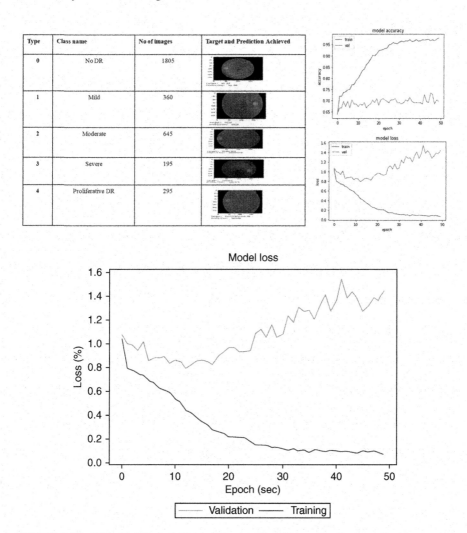

FIGURE 8.10 The classification of deep learning algorithms using fundus retinal images.

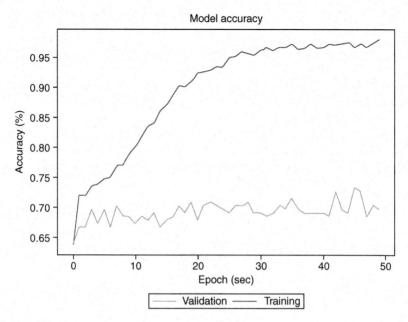

FIGURE 8.10 Continued

As a result, changes in the diabetic retinal image help in deciding the severity of the illness. Hence, DR distinctly varies and could be used for better study in demonstrating differences between no DR, mild, moderate, and severity subjects, as shown in Figure 8.11. From these observations, it was shown that the retinal fundus image of a diabetic person effectively demarcates the clinical pathology.

8.6   CONCLUSION

DL is a fast-growing concept that functions similarly to the human mind. It can effectively overcome the selectivity-invariance problem by representing data at several levels. In medical prognosis, DL methods can be employed in many ways. DR is a major impediment of DM that involves gradual retinal degeneration and, in extreme situations, blindness. It is helpful to understand and treat it early to prevent progression and retinal damage. The use of DL to diagnose DR has grown in popularity in recent years, and the current state of research on the use of DL in the diagnosis of DR is presented in this article. As more DL systems mature and become integrated into clinical practice, physicians will be able to treat patients in need more successfully and efficiently.

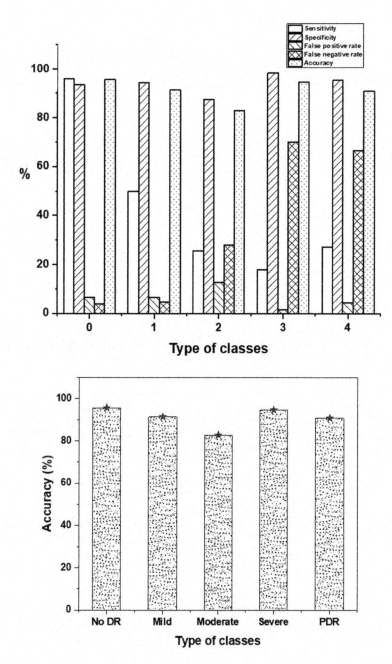

FIGURE 8.11   Accuracy of deep learning algorithms using fundus retinal images.

REFERENCES

[1]   P. Saeedi, I. Petersohn, P. Salpea, and B. Malanda, "Withdrawn: Global and regional diabetes prevalence estimates for 2019 and projections for 2030 and 2045: results from the International Diabetes Federation Diabetes Atlas, 9th edition," *Diabetes Research and Clinical Practice*, p. 107843, 2019, doi: 10.1016/j.diabres.2019.107843.

[2]   N. Nanayakkara *et al.*, "Impact of age at type 2 diabetes mellitus diagnosis on mortality and vascular complications: systematic review and meta-analyses," *Diabetologia*, vol. 64, no. 2, pp. 275–287, 2020.

[3]   D. C. Klonoff, "Benefits and Limitations of Self-Monitoring of Blood Glucose," *Journal of Diabetes Science and Technology*, vol. 1, no.1, pp. 130–132, 2007.

[4]   R. Patterson *et al.*, "Sedentary behaviour and risk of all-cause, cardiovascular and cancer mortality, and incident type 2 diabetes: a systematic review and dose response meta-analysis," *European Journal of Epidemiology*, vol. 33, no. 9, pp. 811–829, 2018, doi: 10.1007/s10654-018-0380-1.

[5]   K. J. Basile, M. E. Johnson, Q. Xia, and S. F. A. Grant, "Genetic susceptibility to type 2 diabetes and obesity: Follow-up of findings from genome-wide association studies," *International Journal of Endocrinology*, 2014, 1–13.

[6]   T. A. Buchanan, A. H. Xiang, and K. A. Page, "Gestational diabetes mellitus: risks and management during and after pregnancy," *National Reviews Endocrinology*, vol. 8, no. 11, pp. 639–649, 2012, doi: 10.1038/nrendo.2012.96.

[7]   S. Lim, J. H. Bae, H. S. Kwon, and M. A. Nauck, "COVID-19 and diabetes mellitus: from pathophysiology to clinical management," *Nature Reviews Endocrinology*, vol. 17, no. 1, pp. 11–30, 2021, doi: 10.1038/s41574-020-00435-4.

[8]   S. A. Lauer *et al.*, "The incubation period of coronavirus disease 2019 (COVID-19) from publicly reported confirmed cases: estimation and application," *Annals of Internal Medicine*, vol. 172, no. 9, pp. 577– 582, 2020, doi: 10.7326/M20-0504.

[9]   M. Gusev *et al.*, "Noninvasive glucose measurement using machine learning and neural network methods and correlation with heart rate variability," *Journal of Sensors*, 2020, doi: 10.1155/2020/9628281.

[10] C. V. Ugwueze, B. C. Ezeokpo, B. I. Nnolim, E. A. Agim, N. C. Anikpo, and K. E. Onyekachi, "COVID-19 and diabetes mellitus: the link and clinical implications," *Dubai Diabetes and Endocrinology Journal*, vol. 26, no. 2, pp. 69–77, 2020, doi: 10.1159/000511354.

[11] K. Dhatariya, "Blood ketones: measurement, interpretation, limitations, and utility in the management of diabetic ketoacidosis," *Review of Diabetic Studies*, vol. 2, pp. 217–225, 2016, doi: 10.1900/RDS.2016.13.217.

[12] S. K. Kermani, A. Khatony, R. Jalali, M. Rezaei, and A. Abdi, "Accuracy and precision of measured blood sugar values by three glucometers compared to the standard technique," *Journal of Clinical and Diagnostic Research*, vol. 11, no. 4, pp. OC05-OC08, 2017, doi: 10.7860/JCDR/2017/23926.9613.

[13] A. S. Lundervold and A. Lundervold, "An overview of deep learning in medical imaging focusing on MRI," *Zeitschrift für Medizinische Physik*, vol. 29, no. 2, pp. 102–127, 2019, doi: 10.1016/j.zemedi.2018.11.002.

[14] F. Girard, C. Kavalec, and F. Cheriet, "Joint segmentation and classification of retinal arteries/veins from fundus images," *Artificial Intelligence in Medicine*, vol. 94, pp. 96–109, 2019, doi: 10.1016/j.artmed.2019.02.004.

[15] B. B. Lahiri, S. Bagavathiappan, T. Jayakumar, and J. Philip, "Medical applications of infrared thermography: a review," *Infrared Physics and Technology*, vol. 55, no. 4, pp. 221–235, 2012, doi: 10.1016/j.infrared.2012.03.007.

[16] M. T. Esfahani, M. Ghaderi, and R. Kafiyeh, "Classification of diabetic and normal fundus images using new deep learning method," *Leonardo Electronic Journal of Practices and Technologies*, vol. 32, pp. 233–248, 2018.

[17] B. M. Leon and T. M. Maddox, "Diabetes and cardiovascular disease: epidemiology, biological mechanisms, treatment recommendations and future research," *World Journal of Diabetes*, vol. 6, no. 13, pp. 1246–1258, 2015, doi: 10.4239/wjd. v6.i13.1246.

[18] O. Y. Atkov *et al.*, "Coronary heart disease diagnosis by artificial neural networks including genetic polymorphisms and clinical parameters," *Journal of Cardiology*, vol. 59, no. 2, pp. 190–194, 2012, doi: 10.1016/j.jjcc.2011.11.005.

[19] A. Borai, C. Livingstone, F. Abdelaal, A. Bawazeer, V. Keti, and G. Ferns, "The relationship between glycosylated haemoglobin (HbA1c) and measures of insulin resistance across a range of glucose tolerance," *Scandinavian Journal of Clinical and Laboratory Investigation*, vol. 71, no. 2. pp. 168–172, 2011, doi: 10.3109/ 00365513.2010.547947.

[20] W. L. Alyoubi, W. M. Shalash, and M. F. Abulkhair, "Informatics in medicine unlocked diabetic retinopathy detection through deep learning techniques: a review," *Informatics Med. Unlocked*, vol. 20, p. 100377, 2020, doi: 10.1016/j.imu.2020.100377.

[21] F. Huang *et al.*, "Reliability of using retinal vascular fractal dimension as a biomarker in the diabetic retinopathy detection," *Journal of Ophthalmology*, 2016, doi: 10.1155/2016/6259047.

[22] S. Das, S. Pal, and M. Mitra, "Significance of exhaled breath test in clinical aiagnosis: a special focus on the detection of diabetes mellitus," *Journal of Medical and Biological Engineering*, 2016, doi: 10.1007/s40846-016-0164-6.

[23] A. Chanmugam *et al.*, "Relative temperature maximum in wound infection and inflammation as compared with a control subject using long-wave infrared thermography," *Advanced Skin Wound Care*, vol. 30, no. 9, pp. 406–414, 2017.

[24] P. Vashist, S. Singh, N. Gupta, and R. Saxena, "Role of early screening for diabetic retinopathy in patients with diabetes mellitus: an overview," *Indian Journal of Community Medicine*, vol. 36, no. 4, pp. 247– 252, 2011, doi: 10.4103/ 0970-0218.91324.

[25] D. W. Safitri and D. Juniati, "Classification of diabetic retinopathy using fractal dimension analysis of eye fundus image," in *AIP Conference Proceedings*, vol. 1867, August, 2017, doi: 10.1063/1.4994414.

[26] Y. Fujiwara, T. Inukai, Y. Aso, and Y. Takemura, "Thermographic measurement of skin temperature recovery time of extremites in patients with type 2 diabetes mellitus," *Experimental and Clinical Endocrinology and Diabetes*, vol. 108, pp. 463– 469, 2000.

[27] A. Kirimtat, O. Krejcar, A. Selamat, and E. Herrera-Viedma, "FLIR vs SEEK thermal cameras in biomedicine: comparative diagnosis through infrared thermography," *BMC Bioinformatics*, vol. 21, no. 88, pp. 1–10, 2020, doi: 10.1186/ s12859-020-3355-7.

[28] N. J. W. Rattray, Z. Hamrang, D. K. Trivedi, R. Goodacre, and S. J. Fowler, "Taking your breath away: metabolomics breathes life in to personalized medicine," *Trends in Biotechnology*, vol. 32, no. 10, pp. 538–548, 2014, doi: 10.1016/j.tibtech.2014.08.003.

[29] D. J. Beale *et al.*, "A review of analytical techniques and their application in disease diagnosis in breathomics and salivaomics research," *International Journal of Molecular Sciences*, vol. 18, no. 1, 2017, doi: 10.3390/ijms18010024.

[30] B. De Lacy Costello *et al.*, "A review of the volatiles from the healthy human body," *Journal of Breath Research*, vol. 8, no. 1, 2014, doi: 10.1088/1752-7155/8/1/014001.

[31] R. Varma *et al.*, "Prevalence of and risk factors for diabetic macular edema in the United States," *JAMA Ophthalmology*, vol. 132, no. 11, pp. 1334–1340, 2014, doi: 10.1001/jamaophthalmol.2014.2854.

[32] H. N. Mhaskar, S. V. Pereverzyev, and M. D. van der Walt, "A deep learning approach to diabetic blood glucose prediction," *Frontiers in Applied Mathematics and Statistics*, vol. 3, 2017, doi: 10.3389/fams.2017.00014.

[33] J. Shapey *et al.*, "Intraoperative multispectral and hyperspectral label-free imaging: a systematic review of in vivo clinical studies," *Journal of Biophotonics*, 2019, doi: 10.1002/jbio.201800455.

[34] F. Li, C. Wang, X. Liu, Y. Peng, and S. Jin, "A composite model of wound segmentation based on traditional methods and deep neural networks," *Computational Intelligence and Neuroscience*, 2018, doi: 10.1155/2018/4149103.

[35] X. Li, X. Hu, L. Yu, L. Zhu, C.-W. Fu, and P.-A. Heng, "CANet: Cross-disease attention network for joint diabetic retinopathy and diabetic macular edema grading," *IEEE Transactions on Medical Imaging*, vol. 35, no. 5, pp. 1483–1493, 2019, doi: 10.1109/TMI.2019.2951844.

[36] B. Cassidy *et al.*, "DFUC 2020: Analysis towards diabetic foot ulcer detection," *European Journal of Endocrinology*, vol. 17, no. 1, pp. 5–11, 2021.

[37] C. Xiao, E. Choi, and J. Sun, "Opportunities and challenges in developing deep learning models using electronic health records data: a systematic review," *Journal of the American Medical Informatics Association*, vol. 25, no. 10, pp. 1419–1428, 2018, doi: 10.1093/jamia/ocy068.

[38] V. Gulshan *et al.*, "Development and validation of a deep learning algorithm for detection of diabetic retinopathy in retinal fundus photographs," *JAMA*, vol. 94043, pp. 1–9, 2016, doi: 10.1001/jama.2016.17216.

[39] X. Wang, Y. Lu, Y. Wang, and W. Chen, "Diabetic retinopathy stage classification using convolutional neural networks," *2018 IEEE International Conference on Information Reuse and Integration*, pp. 465–471, 2018, doi: 10.1109/IRI.2018.00074.

[40] M. D. Abr *et al.*, "Improved Automated Detection of Diabetic Retinopathy on a Publicly Available Dataset Through Integration of Deep Learning," *IOVS*, vol. 57, no. 13, doi: 10.1167/iovs.16-19964.

[41] T. Shanthi and R. S. Sabeenian, "Modified Alexnet architecture for classification of diabetic retinopathy images R," *Computers and Electrical Engineering*, vol. 76, pp. 56–64, 2019, doi: 10.1016/j.compeleceng.2019.03.004.

[42] E. O. Olaniyi, O. K. Oyedotun, and K. Adnan, "Heart Diseases Diagnosis Using Neural Networks Arbitration," *International Journal of Intelligent Engineering and Systems*, vol. 12, pp. 75–82, 2015, doi: 10.5815/ijisa.2015.12.08.

[43] H. Ayatollahi, L. Gholamhosseini, and M. Salehi, "Predicting coronary artery disease: A comparison between two data mining algorithms," *BMC Public Health*, vol. 19, no. 1, pp. 1–9, 2019, doi: 10.1186/s12889-019-6721-5.

[44] S. Sengupta, A. Singh, H. A. Leopold, T. Gulati, and V. Lakshminarayanan, "Ophthalmic diagnosis using deep learning with fundus images – a critical review," *Artificial Intelligence in Medicine*, vol. 102, p. 101758, 2020, doi: 10.1016/j.artmed.2019.101758.

[45] P. Porwal, S. P. Id, R. K. Id, and M. Kokare, "Indian Diabetic Retinopathy Image Dataset (IDRiD): A database for diabetic retinopathy screening research," *Data*, vol. 3, no. 25, pp. 1–8, 2018, doi: 10.3390/data3030025.

[46] K. Alexiadou, and J. Doupis. "Management of diabetic foot ulcers." *Diabetes Therapy*, vol. 3, 2012, doi: 10.1007/s13300-012-0004-9,

[47] S. Bandalakunta Gururajarao, U. Venkatappa, J. M. Shivaram, M. Y., Sikkandar, and A. Al Amoudi, A. "Infrared thermography and soft computing for diabetic foot assessment." *Machine Learning in Bio-Signal Analysis and Diagnostic Imaging*, 2019, 73–97. https://doi.org/10.1016/B978-0-12-816086-2.00004-7

[48] Y.P. Liu, Z. Li, X. Xu, L. Cong. J. Li, and R. Liang. "Referable diabetic retinopathy identification from eye fundus images with weighted path for convolutional neural network." *Artificial Intelligence in Medicine*, 2019, S0933365718307747–. doi:10.1016/j.artmed.2019.07.002

[49] P. Indoria, and Y. K. Rathore (2018). "A survey: Detection and prediction of diabetes using machine learning techniques." *International Journal of Engineering Research and Technology*, 2018, 7.

Index